D1033106

Risk Assessment of Radon in
Drinking Water

Committee on Risk Assessment of Exposure to Radon in Drinking Water

Board on Radiation Effects Research

Commission on Life Sciences

National Research Council

NATIONAL ACADEMY PRESS
Washington, D.C.   1999

NATIONAL ACADEMY PRESS • 2101 Constitution Avenue, NW • Washington, D.C. 20418

NOTICE: The project that is the subject of this report was approved by the Governing Board of the National Research Council, whose members are drawn from the councils of the National Academy of Sciences, the National Academy of Engineering, and the Institute of Medicine. The members of the committee responsible for the report were chosen for their special competences and with regard to appropriate balance.

This report was prepared under EPA Contract EPA X825492-01-0 between the National Academy of Sciences and the Environmental Protection Agency.

Library of Congress Cataloging-in-Publication Data

Risk assessment of radon in drinking water / Committee on Risk Assessment of Exposure to Radon in Drinking Water, Board on Radiation Effects Research, Commission on Life Sciences, National Research Council.
    p. cm.
    Includes bibliographical references and index.
    ISBN 0-309-06292-6 (casebound).
    1. Drinking water—Contamination—United States. 2. Radon—Health aspects. 3. Indoor air pollution—Health aspects—United States. 4. Radon mitigation. 5. Health risk assessment—United States. I. National Research Council (U.S.). Committee on Risk Assessment of Exposure to Radon in Drinking Water.
    RA592.A1 R57 1999   99-6134
    615.9'02—dc21

Risk Assessment of Radon in Drinking Water is available for sale from the National Academy Press, 2101 Constitution Avenue, N.W., Box 285, Washington, DC 20055; 1-800-624-6242 or 202-334-3313 (in the Washington metropolitan area); Internet, http://www.nap.edu

Printed in the United States of America

The National Academy of Sciences is a private, nonprofit, self-perpetuating society of distinguished scholars engaged in scientific and engineering research, dedicated to the furtherance of science and technology and to their use for the general welfare. Upon the authority of the charter granted to it by the Congress in 1863, the Academy has a mandate that requires it to advise the federal government on scientific and technical matters. Dr. Bruce M. Alberts is president of the National Academy of Sciences.

The National Academy of Engineering was established in 1964, under the charter of the National Academy of Sciences, as a parallel organization of outstanding engineers. It is autonomous in its administration and in the selection of its members, sharing with the National Academy of Sciences the responsibility for advising the federal government. The National Academy of Engineering also sponsors engineering programs aimed at meeting national needs, encourages education and research, and recognizes the superior achievements of engineers. Dr. William A. Wulf is the president of the National Academy of Engineering.

The Institute of Medicine was established in 1970 by the National Academy of Sciences to secure the services of eminent members of appropriate professions in the examination of policy matters pertaining to the health of the public. The Institute acts under the responsibility given to the National Academy of Sciences by its congressional charter to be an adviser to the federal government and, upon its own initiative, to identify issues of medical care, research, and education. Dr. Kenneth I. Shine is president of the Institute of Medicine.

The National Research Council was organized by the National Academy of Sciences in 1916 to associate the broad community of science and technology with the Academy's purposes of furthering knowledge and advising the federal government. Functioning in accordance with general policies determined by the Academy, the Council has become the principal operating agency of both the National Academy of Sciences and the National Academy of Engineering in providing services to the government, the public, and the scientific and engineering communities. The Council is administered jointly by both Academies and the Institute of Medicine. Dr. Bruce M. Alberts and Dr. William A. Wulf are chairman and vice chairman, respectively, of the National Research Council.

Preface

At the request of the Environmental Protection Agency (EPA) pursuant to a congressional mandate (amendment to bill S. 1316 to amend title XIV of the Public Health Service Act commonly known as the Safe Drinking Water Act), the National Research Council has appointed a multidisciplinary committee to conduct a study and report on the health risks associated with exposure to radon in drinking water. The committee was also asked to prepare an assessment of the health-risk reduction associated with various mitigation measures to reduce radon in indoor air; to accomplish this task, the committee used the results of the latest scientific studies of risk assessment and relevant peer-reviewed research carried out by organizations and individual investigators. Finally, the committee was asked to summarize the agreements and differences between the various advisory organizations on the issues relevant to the health risks posed by radon in drinking water and radon-mitigation measures and to evaluate the technical and scientific bases of any differences that exist.

The Committee on Risk Assessment of Radon in Drinking Water was appointed in May 1997, held its first meeting on July 14-15, 1997, and held six additional meetings during the next 9 months. The ability of the committee to comply with this extremely tight schedule is a reflection of the dedication and expertise of the committee members and the efforts of the committee staff.

The committee acknowledges the help of those individuals or organizations who gave presentations during our meetings and/or provided information in response to requests by committee members or staff and to others who helped the committee in the completion of our task.

Gustav Akerbloom, Swedish Radiation Protection Institute
Hannu Arvela, Finnish Radiation/Nuclear Safety Authority
Timothy Barry, Environmental Protection Agency
David S. Chase, New Hampshire Radiologic Health Bureau
Gail Charnley, Presidential/Congressional Commission on Risk Assessment
    and Risk Management
Nancy Chiu, Environmental Protection Agency
Jack Correia, Massachusetts General Hospital
Bill Diamond, Environmental Protection Agency
Joe Drago, Kennedy Jenks, San Francisco, CA
Susumo Ito, Professor Emeritus, Harvard University
Dan Krewski, Environmental Health Centre, Ottawa, Canada
Jay Lubin, National Cancer Institute
J.P. Malley, Jr., University of New Hampshire
Sylvia Malm, Environmental Protection Agency
Frank Marcinowski, Environmental Protection Agency
Lars Mjones, Swedish Radiation Protection Institute
Roger McClellan, Chemical Industry Institute of Toxicology
Neal S. Nelson, Environmental Protection Agency
David Paris, Waterworks, Manchester, NH
Dan Pederson, American Water Works Association
Frederick Pontius, American Water Works Association
Jerome Puskin, Environmental Protection Agency
Edith Robbins, New York University
David Rowson, Environmental Protection Agency
Richard Toohey, Oak Ridge Institute of Science and Education
George Sachs, VA Medical Center, Los Angeles
Anita Schmidt, Environmental Protection Agency
Daniel J. Steck, St. John's University
Grant Stemmerman, University of Cincinnati
Neil Weinstein, Rutgers University
Jeanette Wiltse, Environmental Protection Agency

This report has been reviewed by individuals chosen for their diverse per-spectives and technical expertise, in accordance with procedures approved by the National Research Council's Report Review Committee. The purpose of this independent review is to provide candid and critical comments that will assist the authors and the National Research Council in making their published report as sound as possible and to ensure that the report meets institutional standards for objectivity, evidence, and responsiveness to the study charge. The content of the review comments and draft manuscript remain confidential to protect the integrity of the deliberative process. We wish to thank the following individuals for their participation in the review of this report:

Antone Brooks, Washington State University, Tri-Cities
Bernard Cohen, University of Pittsburgh
Douglas Crawford-Brown, University of North Carolina-Chapel Hill
Robert E. Forster, The University of Pennsylvania School of Medicine
Sharon Friedman, Lehigh University
Patricia L. Gardner, New Jersey Department of Environmental Protection
Roger O. McClellan, Chemical Industry Institute of Toxicology
Gilbert Omenn, University of Washington
Frank H. Stillinger, Bell Laboratories
Rhodes Trussell, Montgomery Watson, Inc.

The committee members would like to express their gratitude to the staff of the National Research Council's Board on Radiation Effects Research. The committee members are especially appreciative for study director Steven Simon's technical guidance and encouragement. They are also grateful to Karen Bryant and Doris Taylor for assistance with administrative details related to the committee's work.

Contents

Public Summary

Radiation is a natural part of the environment in which we live. All people receive exposure from naturally occurring radioactivity in soil, water, air and food. The largest fraction of the natural radiation exposure we receive comes from a radioactive gas, radon. Radon is emitted from uranium, a naturally occurring mineral in rocks and soil; thus, radon is present virtually everywhere on the earth, but particularly over land. Thus, low levels of radon are present in all the air we breathe. There are three forms of radon, but the use of the term *radon* in this report refers specifically to radon-222. Although it cannot be detected by a person's senses, radon and its radioactive by-products are a health concern because they can cause lung cancer when inhaled over many years. A recent report by the National Research Council suggested that between 3,000 and 32,000 lung-cancer deaths each year (the most likely value is given as 19,000 deaths) in the United States are associated with breathing radon and its radioactive by-products in indoor air, but these deaths are mainly among people who also smoke.

Most of the radon that enters a building comes directly from soil that is in contact with or beneath the basement or foundation. Radon is also found in well water and will enter a home whenever this water is used. In many situations such as showering, washing clothes, and flushing toilets, radon is released from the water and mixes with the indoor air. Thus, radon from water contributes to the total inhalation risk associated with radon in indoor air. In addition to this, drinking water contains dissolved radon and the radiation emitted by radon and its radioactive decay products exposes sensitive cells in the stomach as well as other organs once it is absorbed into the bloodstream. This report examines to what degree this ingested radon is a health risk and to what extent radon released from water into air increases the health risk due to radon already in the air in homes.

Approximately half of the drinking water in the United States comes from ground water that is tapped by wells. Underground, this water often moves through rock containing natural uranium that releases radon to the water. Water from wells normally has much higher concentrations of radon than does surface water such as lakes and streams.

Radon concentrations can be measured either in terms of a volume of air (becquerel of radon per cubic meter) or a volume of water (becquerel of radon per liter). The average concentration of radon in public water supplies derived from ground water sources is about 20 becquerel per liter (540 pCi). Some wells have been identified with high concentrations, up to 400 times the average. Surface water, such as in lakes and streams, has the lowest concentrations, about one-tenth that of most wells.

Drinking-water quality in the United States is regulated by the Environmental Protection Agency (EPA) under the Safe Drinking Water Act (SDWA). Since radon is acknowledged as a cancer-causing substance, the law directs EPA to set a maximum contaminant level (MCL) for radon to restrict the exposure of the public to the extent that is possible, that is, as close to zero as is feasible.

In 1991, EPA proposed an MCL for radon of 11 becquerel per liter (about 300 pCi per liter) for radon in drinking water. In 2000, the agency is required to set a new MCL based in part on this report. The law also directed EPA to set an alternative MCL (AMCL); an AMCL is the concentration of radon in water that would cause an increase of radon in indoor air that is no greater than the level of radon naturally present in outdoor air. Limiting public risk from radon by treating the water alone is not feasible because radon is also naturally present in the air. Thus, the AMCL is the tool that allows EPA to limit exposure to radon in water to a practical level, that is, allowing no more risk from the radon in water than is posed by the level of radon naturally present in outdoor air.

The 1996 amendments to the Safe Drinking Water Act required EPA to fund the National Academy of Sciences (NAS) to determine the risk from radon in drinking water and also to determine the public-health benefits of various methods of removing radon from indoor air.

In response to that agreement, the NAS established through its principal operating agency, the National Research Council, a committee which has evaluated various issues related to the risk from radon in drinking water and provides here the information needed by EPA to set the AMCL. The primary conclusion from the committee's investigation into the risk of inhaling radon as compared to drinking water containing dissolved radon is as follows:

> Most of the cancer risk resulting from radon in the household water supply is due to inhalation of the radioactive by-products that are produced from radon that has been released from the water into the air, rather than from drinking the water. (The risk from radon is higher among smokers because the combination

of radon and smoke has a greater damaging effect than the sum of the individual risks.) Furthermore, the increased level of indoor radon that is caused by using water in the home is generally small compared with the level of indoor radon that originated in the soil beneath the home.

Based on an analysis of the available data on radon concentrations outdoors and on the transfer from water to air, the Research Council committee arrived at these additional conclusions:

• The average outdoor air concentration over the entire United States is about 15 becquerel per cubic meter (405 pCi per cubic meter or 0.4 pCi per liter).

• The contribution to radon concentration in indoor air from household usage of water is very low—only about one ten-thousandth the water concentration. The reason the resulting airborne concentration is so low is because only about half of the radon in the household water supply escapes into the air and then it is diluted into the large volume of air inside the home.

• Combining this information, the committee has determined that the level of radon in drinking water that would cause an increase of radon in indoor air that is no greater than the level of radon naturally present in outdoor air is about 150 becquerel per liter (4,050 pCi per liter). This conclusion will affect the public and water utilities in the following ways:

1. People who own their own wells are not legally obliged to do anything because the Safe Drinking Water Act does not regulate private wells. However, people who are served by private wells and who wish to minimize their risk should test their water and consider taking action to reduce the radon if the concentration in the water is above the AMCL. In addition, those people should also measure the indoor air concentration in their home and consider taking actions to reduce it if it is above EPA's recommended action levels. Lastly, as the earlier NRC report concluded, stopping smoking is the most effective way to reduce the risk of lung cancer and reduce the risks associated with radon.

2. Water supplies serving 25 or more people or with 15 or more connections are considered to be public water supplies. Those supplies, along with some special cases such as schools, will be subject to radon regulation if they rely on groundwater. In this case, there are three possibilities: (a) The radon in the water is already below the MCL. This will apply to the majority of people in the United States—only about 1 of every 14 individuals routinely consumes water with concentrations greater than the 1991 proposed MCL (11 becquerel per liter or 300 pCi/L). For water below the MCL, nothing needs be done. (b) The radon in the water is greater than the AMCL. In this case, radon reduction (mitigation) would be required by law after the regulation is final. Data available to the committee indicate that there are several types of water mitigation technology that could effectively reduce the radon concentration to the MCL. (c) The radon in the water is between the MCL and the AMCL. In this case, the concentration

must be reduced to the MCL or, if there is an approved state plan, the risk to the population served by the water supply can be reduced by activities that reduce radon in air and/or water.

The committee discussed a variety of methods to reduce radon entry into homes and the concentrations in the indoor air and in water. Ventilation systems can be used to reduce radon concentrations in indoor air to acceptable levels. Periodic testing would be needed to ensure the continued successful operation of individual air treatment systems. New homes can be constructed using methods to reduce airborne radon (radon resistant construction). However, there is not enough evidence at the present time to be certain these techniques are effective. Several water-treatment technologies to remove radon from water are very effective, however, they do not address the largest risk to the occupants of the house, namely radon in air.

The EPA mandate is to reduce public risk caused by exposure to radon. For those communities where the public water supply contains radon at concentrations between the MCL and the AMCL, the law will allow individual states to reduce the risk to their population through multimedia measures to mitigate radon levels in indoor air. A state may develop and submit a multimedia program to mitigate radon levels in indoor air for approval by the EPA Administrator. The Administrator shall approve a state program if the health risk reduction benefits expected to be achieved by the program are equal to or greater than the health risk reduction benefits that would be achieved if each public water system in the state complied with the MCL. If the program is approved, public water systems in the state may comply with the alternative maximum contaminant level in lieu of the MCL. State programs may rely on a variety of mitigation measures, including public education, home radon testing, training, technical assistance, remediation grant and loan or other financial incentive programs, or other regulatory or nonregulatory measures. As required by SDWA, EPA is developing guidelines for multimedia mitigation programs. If there is no approved state multimedia mitigation program, any public water system in the state may submit a program for approval by the EPA Administrator, according to the same criteria, conditions and approval process that would apply to a state program. In this scenario, water utilities can minimize the level of risk to their consumers—even if the water they provide is higher than the MCL (but lower than the AMCL)—by reducing airborne radon in some of the community's homes. Because the risk caused by inhaled radon is so much greater than that caused by radon that is swallowed in water, reducing the airborne radon in only a few homes may reduce public risk enough for the water utility to be in compliance with the multimedia program requirements.

With regard to multimedia programs, the committee's report provides discussion of risk-reduction methods at the community level and of ways to evaluate the effectiveness of reducing radon-related risk within a community or region served by a water utility. One risk reduction technique is public education programs to encourage radon mitigation from indoor air. The previously conducted education and outreach programs reviewed by the committee were largely unsuccessful; therefore, the committee concluded that public education and outreach programs alone would be insufficient to achieve a measurable reduction in health risk.

A multimedia mitigation program will reduce radon risks in indoor air in lieu of reduction to the MCL in drinking water. The specific design of each community water utility's program will depend on many factors. At the same time, complicated risk-reduction programs like those discussed here have many potential difficulties. For example, for water utilities that provide water that contains radon at levels between the MCL and the AMCL, the feasibility of using a multimedia mitigation program will depend on whether there are homes with relatively high indoor radon concentrations. Only in those homes is it feasible to reduce the air concentration sufficiently such that an expensive, large-scale water mitigation program in the region is not needed to satisfy the multimedia program requirements. The key issue is determining how many buildings must have air mitigation systems to obtain a reduction in public risk equal to that which could be achieved by reducing radon in the water supplied to the community. Moreover, air monitoring programs will be needed to identify the homes whose indoor air must be mitigated and effective outreach programs will be needed to educate the public about the need to modify these homes to reduce indoor radon so that the water utility can demonstrate the risk reduction needed for compliance. Finally, consideration needs to be given to how the costs of mitigation of private homes will be apportioned among homeowners and the water utilities or state government.

Another potential problem is the present-day scarcity of trained personnel (particularly in the water utilities) that could design or maintain home air mitigation systems and carry out the tests needed to ensure continued performance of these systems.

Finally, the committee recognizes that the reduction in risk by multimedia programs will not be distributed equally among the public. The mitigation of indoor-air radon in a small number of homes means risk reduction among only a few people who had high initial risk, rather than a uniform risk reduction for a whole population served by the water utility.

The various analyses conducted allowed the committee to estimate the risk and annual number of fatalities caused by radon in water and to compare it with the risk caused by radon in air. The figure presented here summarizes the cancer risk posed by inhaling radon in air (with and without the addition of radon from using water in the home) and the risk posed by drinking water that contains dissolved radon. Specifically, in 1998 in the United States, there will be about 160,000 deaths from lung cancer, mainly as a result of smoking tobacco. Of those, about 19,000 are estimated to result from inhaling radon gas in the home;

though most of these deaths will be among people who smoke. Of the 19,000 deaths, only 160 are estimated to result from inhaling radon that was emitted from water used in the home though most of these deaths would also among smokers. As a benchmark for comparison, about 700 lung-cancer deaths each year can be attributed to exposure to natural levels of radon while people are outdoors.

The committee determined that the risk of stomach cancer caused by drinking water that contains dissolved radon is extremely small and would probably result in about 20 deaths annually compared with the 13,000 deaths from stomach cancer that arises from other causes.

Except in situations where concentrations of radon in water are very high, reducing the radon in water will generally not make a large difference in the total radon-related health risks to occupants of dwellings. Using techniques to reduce airborne radon and its related lung-cancer risk makes good sense from a public-health perspective. However, there are concerns about the equity of the multimedia approach.

The committee concludes that evaluating whether a multimedia approach to radon reduction will achieve an acceptable risk reduction in a cost-effective and equitable manner will be a complex process. It will require significant cooperation among EPA, state agencies, water utilities and local governments, especially because many of the communities affected by the radon regulation will be very small and they will need assistance in making decisions concerning the advantages or disadvantages of a multimedia program. Thus, each public water supply will find it necessary to study its own circumstances carefully before deciding to undertake a multimedia mitigation program instead of treating the water to reduce the radon dissolved in it.

COMPARISON OF LUNG AND STOMACH CANCER FATALITIES IN THE UNITED STATES IN 1998

CANCER RISK DUE TO RADON RESULTS FROM INHALATION OF AIR AND INGESTION OF WATER

INHALATION RISK
breathing air with radon

INGESTION RISK
drinking water with dissolved radon

160,000 lung cancers,[1] mainly from smoking

Estimated 19,000 lung cancers from breathing[2] radon in indoor air (most deaths among smokers)

14,000 stomach cancer from all causes[1]

Estimated 700 lung cancers from[3] breathing radon outdoors

Estimated 160 lung cancers from breathing radon[3] emitted from water in the home

Estimated 20 stomach cancers from[3] drinking water containing radon

[1] American Cancer Society (1998)
[2] BEIR VI Report (National Research Council 1999)
[3] This report

Executive Summary

BACKGROUND

Of all the radioisotopes that contribute to natural background radiation, radon presents the largest risk to human health. There are three naturally occurring isotopes of radon, but the use of the term *radon* in this report refers specifically to ^{222}Rn, which is a decay product of ^{238}U. A recent report by the National Research Council suggested that between 3,000 and 32,000 lung-cancer deaths annually (the most likely value for the number of deaths is 19,000) in the United States are associated with exposure to ^{222}Rn and its short-lived decay products in indoor air, largely because radon substantially increases the lung-cancer risk for smokers.

Most radon enters homes via migration of soil gas. Throughout this report, radon activity concentrations are cited in the SI[1] unit of becquerel per cubic meter (Bq m^{-3}; 1 Bq m^{-3} = 0.027 pCi L^{-1}). The mean annual radon concentration measured in the living areas of homes in the United States is 46 Bq m^{-3}.

Radon has also been identified as a public-health concern when present in drinking water. Surface waters contain a low concentration of dissolved radon. Typically radon concentrations in surface waters are less than 4,000 Bq m^{-3}.[2] Water from ground water systems can have relatively high levels of dissolved radon, however. Concentrations of 10,000,000 Bq m^{-3} or more are known to exist in public water supplies. Many of the water supplies containing substantial concentrations of radon serve very small communities (<1,000 people). Data on

---

[1] International System of Units (SI) adopted in 1960 by the 11th General Conference on Weights and Measurements (see for example NIST 1995; NIST 1991).

[2] Note that 1 cubic meter (m^3) is equivalent in volume to 1,000 L. Thus, 4,000 Bq m^{-3} is equivalent to 4 Bq L^{-1}.

radon in water from public water supplies indicate that elevated concentrations of radon in water occur primarily in the New England states, the Appalachian states, the Rocky Mountain states, and small areas of the Southwest and the Great Plains.

Because radon is easily released by agitation in water, many uses of water release radon into the indoor air, which contributes to the total indoor airborne radon concentration. Ingestion of radon in water is also thought to pose a direct health risk through irradiation of sensitive cells in the gastrointestinal tract and in other organs once it is absorbed into the bloodstream. Thus, radon in drinking water could potentially produce adverse health effects in addition to lung cancer.

Drinking-water quality in the United States is regulated by the Environmental Protection Agency (EPA) under the Safe Drinking Water Act originally passed in 1974. In the 1986 amendments to the act, EPA was specifically directed to promulgate a standard for radon as one of several radionuclides to be regulated in drinking water. Because of delays in implementing the regulation of radionuclides in drinking water, EPA was sued. In a consent decree, EPA agreed to publish final rules for radionuclides in drinking water, including radon, by April 1993.

EPA proposed national primary drinking water regulations for radionuclides in 1991. Because radon is a known carcinogen, its maximum contaminant level goal (MCLG) was automatically set at zero. A maximum contaminant level (MCL) of 11,000 Bq m^{-3} was subsequently proposed as the level protective of public health and feasible to implement taking costs into account. Public comments on the proposed regulations suggested that the MCL for radon be set somewhere from less than 1,000 Bq m^{-3} to 740,000 Bq m^{-3}; a large majority favored setting the MCL at value higher than 11,000 Bq m^{-3}.

In 1992, Congress directed the Office of Technology Assessment to analyze the EPA health risk assessment and outline actions that could address regulation of radon, considering both air and water. Also in 1992, the Chaffee-Lautenberg amendment to the EPA appropriation bill for FY 1993 directed the agency to seek an extension of the deadline for publishing a final rule until October 1993 and to submit a report, reviewed by EPA's Science Advisory Board (SAB), to Congress by July 1993. That report was to address the risks posed by human exposure to radon and consider both air and water sources, the costs of controlling or mitigating exposure to waterborne radon, and the risks posed by treating water to remove radon. The SAB review of the report questioned EPA's estimates of the number of community water supplies affected, the extrapolation of the risk of lung cancer associated with the high radon exposures of uranium miners to the low levels of exposure experienced in domestic environments and the magnitude of risk associated with ingestion. The SAB report also emphasized that the risk of cancer from radon in domestic settings was a multimedia issue and that the risk for radon in water must be considered within the context of the total risk from radon, which is dominated by radon in indoor air. The Office of Management and

Budget also expressed concern about EPA's analysis of the cost of mitigation. In the agency's FY 1994 appropriation bill, Congress ordered EPA to delay publishing a rule for radon in drinking water.

The 1996 amendments to the Safe Drinking Water Act required EPA to contract with the National Academy of Sciences (NAS) to conduct a risk assessment of radon in drinking water and an assessment of the health-risk reduction benefits associated with various measures to reduce radon concentrations in indoor air. EPA is also required to publish an analysis of the health-risk reduction and the costs associated with compliance with any specific MCL before issuing a proposed regulation. The law also directed EPA to promulgate an alternative maximum contaminant level (AMCL) if the proposed MCL is less than the concentration of radon in water "necessary to reduce the contribution of radon in indoor air from drinking water to a concentration that is equivalent to the national average concentration of radon in outdoor air." Under the law, states may develop a multimedia mitigation progam which if approved by EPA would allow utilities whose water has radon concentrations higher than the MCL, but lower than the AMCL, to comply with the AMCL. The multimedia programs to mitigate radon in indoor air may include "public education; testing; training; technical assistance; remediation grants, loan or incentive programs; or other regulatory or non-regulatory measures." If a state does not have an EPA-approved multimedia mitigation program, a public water supply in that state may submit such a program to EPA directly. Public water supplies exceeding the AMCL and choosing to institute a multimedia mitigation program to achieve equivalent health risk reductions must, at a minimum, treat their water to reduce radon in water concentrations to less than or equal to the AMCL. The present report was written to address the issues just discussed.

CRITICAL ISSUES

It has been difficult to set a standard for radon, as opposed to other radionuclides in drinking water, because of the absence of authoritative dosimetric information for radon dissolved in water. Furthermore, radon presents a unique regulatory problem in that its efficient transfer from water into indoor air produces a risk from the inhalation of its decay products. Thus, it is regulated as a radionuclide in water, but a major portion of the associated risk occurs because of its contribution to the airborne radon concentration.

Because of the relatively small volume of water used in homes, the large volume of air into which the radon is emitted, and the exchange of indoor air with the ambient atmosphere, **radon in water typically adds only a small increment to the indoor air concentration**. Specifically, radon at a given concentration in water adds only about 1/10,000 as much to the air concentration; that is, typical use of water containing radon at 10,000 Bq m^{-3} will on average increase the air radon concentration by only 1 Bq m^{-3}. **There is always radon in indoor air**

from the penetration of soil gas into homes, so only very high concentrations of radon in water will make an important contribution to the airborne concentration.

Even though water generally makes only a small contribution to the indoor airborne radon concentration, the risk posed by radon released from water, even at typical groundwater concentrations, is estimated to be larger than the risks posed by the other drinking water contaminants that have been subjected to regulation, such as disinfection by-products. Thus, in most homes, the risk to the occupants posed by indoor radon is dominated by the radon from soil gas, which is not subject to regulation, and a change in the radon in drinking water would produce a minimal change in the risk posed by airborne radon. This problem led to the suggestion that mitigation of radon in indoor air be considered an alternative means of achieving risk reduction equal to or greater than that which would be achieved by reducing the concentration of radon in drinking water.

The ingestion of radon in water also presents a possible risk. Questions were raised with respect to the ingestion risk assessment that EPA used in the 1991 proposed regulations and in the revised multimedia risk assessment of 1994. The questions were related to the applicability of some of the data used as the basis of the risk model and to the resulting assumptions that were used to estimate risk. The substantial uncertainties in the radon health risks other than those posed by inhalation add to the problems of setting an appropriate MCL to protect public health. Thus, a reevaluation of the ingestion risks was needed.

COMMITTEE CHARGE

EPA contracted with NAS to address the issues cited above, and the committee on the Risk Assessment of Radon Gas in Drinking Water was formed in the National Research Council's Board on Radiation Effects Research. The specific tasks assigned to the committee were:

• To examine the development of radon risk assessments for both inhalation and ingestion of water.
• To modify an existing risk model if that was deemed appropriate or to develop a new one if not.
• To review the scientific data and technical methods used to arrive at risk coefficients for radon in water.
• To assess potential health-risk reduction benefits associated with various mitigation measures to reduce radon in indoor air.

The final report includes:

• Estimates of cancer risk per unit activity concentration of radon in water.
• Assessment of the state of knowledge with respect to health effects of

radon in drinking water for populations at risk, such as infants, children, pregnant women, smokers, elderly persons, and seriously ill persons.

• Review of information regarding teratogenic and reproductive effects in men and women due to radon in water.

• Estimates of the transfer coefficient relating radon in water to average radon concentrations in indoor air.

• Estimates of average radon concentrations in ambient air.

• Estimates of increased health risks that could result from methods used to comply with regulations for radon in drinking water.

• Discussion of health-risk reduction benefits obtained by reducing radon using currently available methods developed for reducing radon concentrations in indoor air and comparison of these benefits with those achievable by the comparable reduction of risks associated with mitigation of radon in water.

FINDINGS AND CONCLUSIONS

The committee's report addresses each of those points, and its conclusions are summarized below. The order of presentation below follows that in the report.

Occurrence of Radon in the United States

National data on indoor radon, radon in water, and geologic radon potential indicate systematic differences in the distribution of radon across the United States. Geologic radon-potential maps and statistical modeling of indoor radon exposures make it clear that the northern United States, the Appalachian and Rocky Mountain states, and states in the glaciated portions of the Great Plains tend to have higher than average indoor radon concentrations. Some smaller areas of the southern states also have higher than average indoor radon concentrations. Data on radon in water from public water supplies indicate that elevated concentrations of radon in water occur in the New England states, the Appalachian states, the Rocky Mountain states, and small areas of the Southwest and the Great Plains.

National Average Ambient Radon Concentration

The ambient concentration of radon varies with distance from and height over its principal source in the ground (rocks and soil) and from other sources that can locally or regionally affect it, such as lakes, mine or mill tailings, vegetation, and fossil-fuel combustion. However, diurnal fluctuations due to changes in air stability and meteorologic events account for most of the variability. Average ambient radon concentrations were measured by EPA over nine seasons at 50 sites across the United States. Most, but not all, sites coincided with the capital city of the state but did not statistically represent the population across the U.S., nor were the measurement at these sites necessarily representative of average ambient radon concentrations in each state. But the EPA data set is the only one with a fully national extent. The committee does not believe that the data are sufficiently representative to provide a population-weighted annual average ambient radon concentration. **An unweighted arithmetic mean radon concentration of 15 Bq m^{-3}, with a standard error of 0.3 Bq m^{-3} was calculated based on the EPA data set, and the committee recommends use of this value as the best available national ambient average concentration.** After reviewing all the other ambient radon concentration data that are available from other specific sites, the committee concluded that the national average ambient radon concentration would lie between 14 and 16 Bq m^{-3}.

Transfer Coefficient

The transfer coefficient is the average fraction of the initial average radon concentration in water that is contributed to the indoor airborne radon concentration. The average transfer coefficient estimated by a model and the average estimated from measurement data are in reasonable agreement. The average of the measurements was 0.9×10^{-4} with a standard error of 0.1×10^{-4}, and the model's average was either 0.9×10^{-4} or 1.2×10^{-4} depending on the choice of input parameter values. **Having considered the problems with both the measurements of the transfer coefficient and the measurements that are the input values into the model, the committee concludes that the transfer coefficient is between 0.8×10^{-4} and 1.2×10^{-4} and recommends that EPA continue to use 1.0×10^{-4} as the best central estimate of the transfer coefficient that can now be obtained.**

Biologic Basis of Risk Estimation

The biologic effects of radon exposure under the low exposure conditions found in domestic environments are postulated to be initiated by the passage of single alpha particles with very high linear energy transfer. The alpha-particle tracks produce multiple sites of DNA damage that result in deletions and rearrangements of chromosomal regions and lead to the genetic instabilities implicated in tumor progression. Because low exposure conditions involve cells exposed to single tracks, variations in exposure translate into variations in the number of exposed cells, rather than in the amount of damage per cell. **This mechanistic interpretation is consistent with a linear, no-threshold relationship between high-linear energy transfer (high-LET) radiation exposure and cancer risk, as was adopted by the BEIR VI committee. However, quantitative estimation of cancer risk requires assumptions about the probability of an exposed cell becoming transformed and the latent period before malignant transformation is complete. When these values are known for singly hit** cells, the results might lead to reconsideration of the linear no-threshold assumption used at present.

Ingestion Risk

The cancer risk arising from ingestion of radon dissolved in water must be derived from calculations of the dose absorbed by the tissues at risk because no studies have quantified the risk. Studies of the behavior of radon and other inert gases have established that they are absorbed from the gastrointestinal tract and readily eliminated from the body through the lungs. The stomach, the portal of entry of ingested radon into the body, is of particular concern. The range of alpha particles emitted by radon and its short-lived decay products is such that alpha particles emitted within the stomach are unable to reach the cells at risk in the stomach wall. Thus, the dose to the wall depends heavily on the extent to which radon diffuses from the contents into the wall. Once radon has entered the blood, through either the stomach or the small intestine, it is distributed among the organs according to the blood flow to them and the relative solubility of radon in the organs and in blood. Radon dissolved in blood that enters the lung will equilibrate with air in the gas-exchange region and be removed from the body.

The committee found it necessary to formulate new mathematical models of the diffusion of radon in the stomach and the behavior of radon dissolved in blood and other tissues. The need for that effort arose from the lack of directly applicable experimental observations and from limitations in the extent to which one can interpret available studies. The diffusion of radon within the stomach was modeled to determine the expected time-integrated concentration of radon at the depth of the cells at risk. The result, based on a diffusion coefficient of 5×10^{-6} cm^2 s^{-1}, indicated that a conservative estimate of the integrated concentration in the wall was about 30% of that in the stomach content.

The committee also found it useful to set forth a physiologically-based pharmacokinetic (PBPK) model of the behavior of radon in the body. Various investigators have assessed the retention of inhaled and ingested radon in the body, but their observations do not relate directly to the distribution of radon among the tissues. The PBPK is formulated using information on blood flow to the tissues and on the relative solubility of radon in blood and tissue to determine the major tissue of deposition (which was adipose tissue) and retention within this tissue. The PBPK model is consistent with the observations regarding radon behavior in the body. Unlike previous estimates of the radiation dose, the committee's analysis also considered that each radioactive decay product formed from radon decay in the body exhibited its own behavior with respect to tissues of deposition, retention, and routes of excretion.

The committee's estimates of cancer risk are based on calculations with risk-projection models for specific cancer sites. The computational method was that described in EPA's *Federal Guidance Report 13*. **An age- and gender-averaged**

cancer death risk from lifetime ingestion of radon dissolved in drinking water at a concentration of 1 Bq m^{-3} is 0.2 × 10^{-8}. Stomach cancer is the major contributor to the risk. The actual risk from ingested radon could be as low as zero depending on the validity of the linear, no-threshold dose response hypothesis, however, the committee has estimated confidence limits on the ingestion risk (see chapter 4).

Inhalation Risk

Lung cancer arising from exposure to radon and its decay products is bronchogenic. The alpha-particle dose delivered to the target cells in the bronchial epithelium is necessarily modeled on the basis of physical and biologic factors. The dose depends particularly on the diameter of the inhaled ambient aerosol particles to which most of the decay products attach. These particles deposit on the airway surfaces and deliver the pertinent dose, and the dose can vary, because of changes in particle size, by about a factor of 2 in normal home conditions.

The dose from radon gas itself is smaller than the dose from decay products on the airways, mainly because of the location of the gas in the airway relative to the target cells—that is, the source-to-target geometry. The dose from radon gas that is soluble in body tissues is also smaller than the decay-product dose. Two of the underground-miner studies showed no statistically significant risk of cancer in organs other than the lung due to inhaled radon and radon decay products. The dosimetry supports that observation, although there is a need to continue the miner observations.

The risk of lung cancer associated with lifetime inhalation of radon in air at a concentration of 1 Bq m^{-3} was estimated on the basis of studies of underground miners. The values were based on risk projections from three follow-up studies: BEIR IV (National Research Council 1988), NIH (1994) and BEIR VI (National Research Council 1999). These three reports used data from 4 to 11 cohorts of underground miners in seven countries and developed risk projections of 1.0 × 10^{-4}, 1.2 × 10^{-4}, and 1.3 × 10^{-4} per unit concentration in air (1 Bq m^{-3}), respectively. The three values were for a mixed population of smokers and nonsmokers. **The value adopted by the committee is the rounded average derived from the two BEIR-VI model results and equals 1.6 × 10^{-4} per Bq m^{-3}.** The lung-cancer risk to smokers is statistically significantly higher than the risk to nonsmokers. **Given the adopted transfer coefficient of 1 × 10^{-4}, the risk of lung cancer (discussed in two reports of the National Research Council and one of the National Institutes of Health) posed by lifetime exposure to radon (^{222}Rn) in water at 1 Bq m^{-3} was calculated to be 1.6 × 10^{-8}.**

Summary of Risk Estimates

The risk estimates developed by the committee for radon in drinking water are summarized in table ES-1. Although the committee was asked to estimate the risks to susceptible populations—such as infants, children, pregnant women, smokers, and elderly and seriously ill persons—there is insufficient scientific information to permit such estimation except for the lung-cancer risk to smokers, which is presented separately in the table. **The adopted lifetime risk of lung cancer for a mixed population of smokers and nonsmokers, men and women, resulting from the air exposure to radon from a waterborne radon concentration of 1 Bq m^{-3} is 1.6 × 10^{-8}. The adopted lifetime risk of stomach cancer for the same water concentration is 0.2 × 10^{-8}; the committee could not make a distinction in ingestion risk for any specifically identified subpopulation other than the differences in gender.**

Figure 1 (see Public Summary) puts the inhalation and ingestion risks into perspective by direct comparison of annual cancer deaths. The number of lung-cancer deaths in the United States is estimated to be 160,100 in 1998 (ACS 1998). Using the average of the two BEIR-VI risk models and adjusting for the 1998 increase in the number of lung-cancer deaths, the committee estimates there will be about 19,000 lung-cancer deaths in 1998 attributable to radon and the combination of radon and smoking. The committee estimated there might be about 20 stomach-cancer deaths in 1998 (with a subjectively determined uncertainty range from 1 to 50 deaths) attributable to the ingestion of radon in drinking water as compared to 13,700 stomach-cancer deaths that are estimated to develop in the United States in 1998 from all causes (ACS 1998). Based on an estimated national mean value of radon in drinking water, the committee estimates 160 lung cancer deaths in 1998 (with a subjectively determined range from 25 to 280 deaths) attributable to indoor radon (in air) resulting from the release of radon from household water. **The committee's analysis indicates that most of the cancer risk posed by radon in drinking water arises from the transfer of radon into indoor air and the subsequent inhalation of the radon decay products, and not from the ingestion of the water.**

TABLE ES-1   Committee Estimate of Lifetime Risk Posed by Exposure to Radon in Drinking Water at 1 Bq m^{-3}

Exposure Pathway	Lifetime risk		
	Male	Female	U.S. Population[a]
Inhalation (ever-smokers)[b]	3.1 × 10^{-8}	2.0 × 10^{-8}	2.6 × 10^{-8}
Inhalation (never-smokers)[b]	0.59 × 10^{-8}	0.4 × 10^{-8}	0.50 × 10^{-8}
Inhalation (population)[b]	2.1 × 10^{-8}	1.2 × 10^{-8}	1.6 × 10^{-8}
Ingestion	0.15 × 10^{-8}	0.23 × 10^{-8}	0.19 × 10^{-8}
Total Risk (inhalation and ingestion)	2.2 × 10^{-8}	1.4 × 10^{-8}	1.8 × 10^{-8}

[a]These rounded values combine the various subpopulations, with appropriate weighting factors taken from the 1990 U.S. Census.
[b]Based on the radon decay product risks of BEIR VI Report (National Research Council 1999) and includes the incremental dose to showering with the uncertainties in these estimates.

The committee was asked to review teratogenic and reproductive risks. **There is no scientific evidence of teratogenic and reproductive risks associated with radon in tissues from either inhalation or ingestion.**

Comparison of the Present Analysis with the Previous EPA Analyses

The committee's analysis results in a modest reduction of the overall risk associated with radon in drinking water compared with the two previous analyses conducted by the EPA. However, the magnitudes of the risks associated with the different exposure pathways are different, as shown in table ES-2. **The committee's analysis estimates that the inhalation pathway accounts for about 89% of the estimated cancer risk and ingestion accounts for 11%. In contrast, EPA's 1994 analysis suggested that inhalation accounted for 47% of the overall risk and ingestion accounted for 53%.**

Based on the committee's analysis, the estimated inhalation risk has increased while the estimated ingestion risk has decreased. The committee did not do any new analysis for the inhalation risk. An average risk value based on three studies: BEIR IV, NIH, and BEIR VI (NRC 1988; Lubin et al. 1994; NRC 1999; respectively) was adopted. The committee did conduct a new analysis of the ingestion risk, based on a model developed for this study. This model reduces the overall ingestion risk factor by about a factor of 5, and suggests that, in contrast with the previous EPA analysis, almost all of the ingestion risk is attributed to the stomach. The estimated ingestion risk factors for various organs are compared in table ES-2.

There are a number of factors underlying the analysis of the risk associated with radon in drinking water, in addition to the lifetime radiation risk factors described above. These include the amount of water ingested, the effective expo-

TABLE ES-2 Comparison of Individual Lifetime Risk Estimates Posed by Radon in Drinking Water at a Concentration of 1 Bq m^{-3}

Exposure Pathway	Committee Analysis[a]	1991 EPA Proposed Rule[b]	1994 Revised EPA Analysis[c]
(A) Radon progeny inhalation[a]	1.6×10^{-8}	1.3×10^{-8}	0.81×10^{-8}
(B) Radon inhalation		0.05×10^{-8}	0.054×10^{-8}
(C) Ingestion	0.2×10^{-8}	0.4×10^{-8}	0.95×10^{-8}
Stomach	1.6×10^{-9}	2.0×10^{-9}	4.9×10^{-9}
Colon	0.059×10^{-9}	0.46×10^{-9}	1.4×10^{-9}
Liver	0.058×10^{-9}	0.33×10^{-9}	0.25×10^{-9}
Lung	0.034×10^{-9}	0.55×10^{-9}	1.2×10^{-9}
General tissue	0.079×10^{-9}	0.61×10^{-9}	1.5×10^{-9}
Total risk (A+B+C)	1.8×10^{-8}	1.8×10^{-8}	1.8×10^{-8}

[a]Total for the U.S. population (averaging across sex and smoking status).
[b]EPA 1991b.
[c]EPA 1994b.

sure duration and the overall water-to-air transfer factor. The EPA reanalysis (EPA 1994b) used a direct tapwater consumption rate of 1 L d^{-1}, an exposure time of 70 y, and assumed that 20% of the radon in the tapwater is released from the water in the process of transferring the water from the tap to the stomach (tapwater is defined as water ingested directly, without agitation or heating). The committee used an age- and gender-specific tapwater usage rate that corresponds to an age- and gender-average rate of 0.6 L d^{-1} and assumed all of the radon remained dissolved in the water during the transfer process. Both the EPA and the committee analyses used a transfer factor of 1×10^{-4} for purposes of estimating the contribution radon dissolved in water makes to the overall indoor air radon concentration.

The estimated number of cancer deaths per year from public exposure to radon are compared in table ES-3. Ranges estimated by this committee are approximate and are based on judgment using the best available information.

Uncertainty Analysis

Estimating potential human exposures to and health effects of radon in drinking water involves the use of large amounts of data and the use of models for projecting relationships outside the range of observed data. The data and models must be used to characterize population behaviors, engineered-system performance, contaminant transport, human contact, and dose-response relationships among populations in different areas, so large variabilities and uncertainties are associated with the resulting risk characterization. The report provides an evaluation of the importance of and methods for addressing the uncertainty and variability that arise in the process of assessing multiple-route exposures to and the health risks associated with radon.

TABLE ES-3   Comparison of estimated cancer deaths per year due to exposure to radon and estimated possible ranges due to uncertainty

Exposure Pathway	Committee Analysis[a]	Revised EPA Analysis[c]
Inhalation of radon progeny in indoor air	18,200[b] (3,000-33,000)	13,600
Inhalation of radon progeny in outdoor air	720 (120-1300)	520
Inhalation of radon progeny derived from the release of radon from drinking water	160 (25-290)[d]	86
Ingestion of radon in drinking water	23 (5-50)	100

[a]Based on the 1998 estimated U.S. population of 270 million.
[b]Based on data from BEIR VI (National Research Council 1999).
[c]Based on a U.S. population of 250 million (EPA 1994b).
[d]Values derived from rescaling the analysis of the EPA-SAB (1994b) report using 1998 population and mortality data and risk estimates from BEIR VI (National Research Council 1999).

The data, scenarios, and models used to represent human exposures to radon in drinking water include at least four important relationships (i) The magnitude of the source-medium concentration, that is, the concentrations of radon in the water supply and in other relevant media, such as ambient air, (ii) the contaminant concentration ratio, which defines how much a source-medium concentration changes as a result of transfers, transformation, partitioning, dilution, and so on before human contact, (iii) the extent of human contact, which describes (often on a body-weight basis) the frequency (days per year) and magnitude (liters per day) of human contact with a potentially contaminated exposure medium (tap water, indoor air, or outdoor air), and (iv) the likelihood of a health effect, such as cancer, associated with a predicted extent of human contact. The latter area of uncertainty includes that of the dose-response model assumed. Uncertainties in modeling the movement of radon with the wall of the stomach (model structure), in the model parameters, and the lack of relevant experimental observations are the critical sources of uncertainty. **The key points discussed included one overarching issue, that being how uncertainty and variability can affect the reliability of the estimates of health effects for any exposure scenario and related control strategies.**

Mitigation of Radon in Air

There has been considerable research on and practical experience with the use of active (mechanical) systems for the control of radon entry into buildings. Use of such systems, when they are properly installed and operating, can typically yield indoor airborne radon concentrations below 150 Bq m^{-3} and can often result in concentrations of about 75 Bq m^{-3}. Although there is considerable experience with the design and installation of active systems, monitoring programs are needed to ensure the continued successful operation of individual active systems. Another possible way to reduce risks associated with exposure to airborne radon is to design and build radon-resistant new buildings. Although the technical potential for building radon-resistant buildings has been demonstrated under some circumstances, the scientific basis for ensuring that it can be done reliably and as a consistent outcome of normal design and construction methods is inadequate. **With the exception of the results in research conducted in Florida, there are no comparative data on which to base estimates of the overall effects of radon-resistant construction methods on reducing concentrations of radon in indoor air radon concentrations.**

Mitigation of Radon in Water

Several water treatment technologies have been used to effectively remove radon from water. However various issues and secondary effects must be addressed in connection with each method, including intermedia pollution (transfer

of radon from water to air) in the case of aeration and the retention of radionuclides (gamma-ray exposure and waste disposal) in the case granular activated
carbon (GAC) adsorption. If water must be treated to meet either the AMCL or
the MCL, disinfection might be required to meet the pending groundwater rule.
In this case, the risk associated with the disinfection byproducts, as estimated by
the committee, will be smaller than the risk reduction gained from radon removal.
The committee has estimated the equivalent gamma dose from a GAC system
designed to remove radon from a public water supply. The dose depends heavily
on the details and geometry of the system and should be predicted with an
extended-source model that can be modified to simulate the actual dimensions of
the treatment units.

Multimedia Approach to Risk Reduction

The 1996 Safe Drinking Water Act Amendments permit EPA to establish an
alternative maximum contamination level (AMCL) if the MCL is low enough so
that the contribution of waterborne radon to the indoor radon concentration is less
than the national average concentration in ambient air. The AMCL is defined
such that the waterborne contribution of radon to the indoor air concentration is
equal to the radon concentration in outdoor air, which is taken to be the national
average ambient radon concentration. In the situations where radon concentrations in water are greater than the MCL but less than the AMCL, states or water
utilities can develop a multi-media approach to health risk reduction. The EPA is
required to publish guidelines including criteria for multimedia approaches to
mitigate radon in indoor air that result in a reduction in risk to the population
living in the area served by a public water supply that contains radon in concentrations greater than the MCL. The committee has examined some of the implementation issues involved in a multimedia mitigation approach through a sequence of scenarios that explore the possible options.

The MCL will be determined by EPA based on a variety of considerations
including their risk assessment, measurement technology, and best available
treatment options and thus, a specific value has not yet been determined. The
ratio of the average ambient radon air concentration to the transfer coefficient
defines the AMCL. **On the basis of the committee's recommended values
for the average ambient radon concentration and the average transfer
coefficient, the AMCL would be 150,000 Bq m^{-3} (about 4,000 pCi L^{-1}).**
Water in excess of the AMCL must be mitigated at least to the AMCL, and
alternative means can then be used to provide a health-risk reduction equivalent
to what would be obtained by mitigation of the water to the MCL. However,
because of the relatively small cost difference between mitigating the water to
the AMCL and to the MCL, the committee believes that in most cases multimedia mitigation programs will probably not be considered for public water supplies with water concentrations in excess of the AMCL. For high radon concen-

tration water, it will generally be most cost-effective to mitigate radon in water to the MCL.

For water supplies with radon concentrations between the MCL and the AMCL, the feasibility of implementing a multimedia mitigation program depends on the availability of homes in which the airborne radon concentration is high (greater than 150 Bq m^{-3}). EPA has divided the country into three regions of different potentials for elevated indoor radon concentration. For water supplies in areas of low indoor air radon potential, it will be difficult to identify and mitigate enough homes to achieve an equivalent or better health-risk reduction by treating the air. For such water supplies, it is unlikely that a public water system's multimedia mitigation program will be practical unless the water concentration of radon is only slightly above the MCL.

In areas of medium and high indoor air radon potential, it is more feasible to mitigate a small number of high-indoor-concentration homes to provide an equivalent health-risk reduction at a cost less than the cost of mitigating the water. In this scenario, the public water supply would have to actively recruit high-indoor-air radon concentration homes and mitigate them. Incentives could perhaps be used to get participation of homeowners in these multimedia programs. In addition, the utilities would have to monitor and maintain the air mitigation systems routinely. This scenario would require water utilities to become involved in air mitigation in individual homes, something with which they are likely to have little experience.

Reduction of radon in indoor air can be an alternative means of reducing overall risks associated with radon. One way to achieve this is to install active (mechanical) systems to reduce radon entry into existing or new houses. Adequate testing (long-term measurements in the living space to reflect actual exposures) will be necessary to determine which existing houses should be mitigated. Routine follow-up measurements will be needed, both to determine the risk reduction achieved by the mitigation and to ensure continued successful operation of the mitigation systems. To ensure that health-risk reductions are at least as great as the reductions that would result from reducing the water radon concentration to the MCL, the number of homes with air mitigation systems should be 10-20% greater than the calculated minimum number of homes. Radon-resistant new construction methods could also be used although the technical and practical bases of their implementation are still poorly developed. Evaluation of the baseline radon exposure would require use of radon-monitoring data from existing houses in the community of interest or estimates of average indoor concentrations based on calculated radon potentials for the region. Careful attention to the follow-up monitoring results would be important, both for determining how much radon reduction has resulted (on the basis of aggregate comparisons) and for determining whether radon persists at unacceptable concentrations.

Various educational and outreach programs reviewed by this committee indicate that, in general, public apathy about the potential risks of exposure to radon has generally remained, despite numerous and sometimes costly public education efforts. Though the evaluation of many of these programs has not been rigorous, on the basis of the reported results, **the committee concludes that an education and outreach program would be insufficient to provide a scientifically sound basis for claiming equivalent health-risk reductions and that an active program of mitigation of homes would be needed to demonstrate health-risk reduction.**

Furthermore, the mitigation of indoor-air radon concentrations in a small number of homes means risk reduction among only a few people who had high initial risk, rather than uniform risk reduction for a whole population served by the water utility. This approach raises questions of equity among the various groups that are being exposed to various levels of risk associated with radon. Equity issues would also result if the airborne-radon risks in one community were traded for the risks in another without a resulting identical or improved public health effect and a commensurate economic benefit to both communities. **Non-economic considerations could play a large role in the evaluation of multimedia mitigation programs and might be the deciding factors in whether to undertake such a program.** In any planning process, a careful program of public education, utilizing experts in risk communication, will be essential to give the public an adequate perspective of the tradeoffs in risks being proposed and of the health and economic costs and benefits that will be produced by the various alternatives.

EPA and the state agencies responsible for water quality will continue to be faced with the problem of the health risks associated with the presence of radon in drinking water. The increment in indoor radon that emanates from the water will generally be small compared with the average concentration of radon already present in the dwellings from other sources. Thus, except in situations where concentrations of radon in water are very high, the reduction of radon in water will generally not make a substantial reduction in the total radon-related health risks to occupants of dwellings served by the water supply. However, the risks associated with the waterborne radon are large in comparison with other regulated contaminants in drinking water. Using mitigation of airborne radon to achieve equivalent or greater health-risk reductions therefore makes good sense from a public-health perspective. However, there are concerns that the equity issues associated with the multimedia approach and other related issues will become important in obtaining agreement by all of the stakeholders. This issue will require each public water supply and the regulatory agency overseeing it to study the circumstances carefully before deciding to implement a multimedia mitigation program in lieu of water treatment.

1

Introduction

THE ORIGINS OF RADON

Naturally occurring radioactivity can be found throughout the earth's crust. Some of these radionuclides decay into stable elements, such as $^{40}K \rightarrow {}^{40}Ar$, $^{14}C \rightarrow {}^{14}N$ and $^{87}Rb \rightarrow {}^{87}Sr$. Others are members of sequences of radioactive decay in which one radionuclide decays into another radionuclide. The three principal such series found in nature originate with ^{238}U, ^{235}U, and ^{232}Th (NCRP 1987a).

The immediate disposition of an atom created in a radioactive series depends on physical and chemical properties of the element and on the surrounding soil or rocks. Many of the elements in the process are metals such as uranium, thorium, polonium, lead, and bismuth or alkaline earths such as radium. These elements vary greatly in solubility depending on ambient physical and chemical conditions and may go into solution or be absorbed onto organic particles or clay minerals. Uranium, radium, and radon are the most mobile, lead and bismuth are only moderately mobile, while thorium and polonium remain relatively immobile.

One of the most abundant sources of naturally occurring radioactivity is the series that begins with ^{238}U, which is illustrated in figure 1.1. The first 14 members in this series collectively emit gamma, beta, and alpha radiation. Because of the arrangement of half-lives and chemical properties, the concentration of radioactivity of the early members of the series is proportional to the concentration of ^{238}U in the earth.

An important deviation happens roughly midway through the ^{238}U series: ^{226}Ra decays by alpha emission, thereby creating ^{222}Rn. In contrast with other

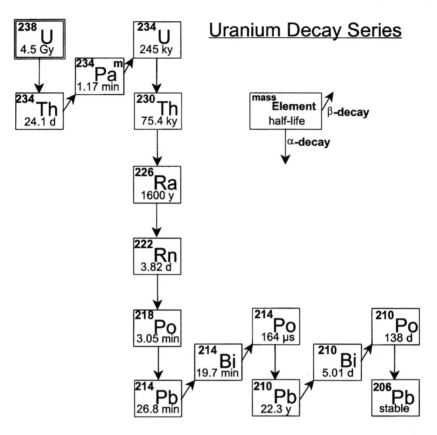

FIGURE 1.1   Decay scheme for natural occurring ^{238}U chain.

members of the series, which are solids, radon is a chemically-inert noble gas and can migrate in the environment.

MIGRATION OF RADON

A radon atom that is created deep within a grain of rock usually remains there until it decays. However, when a radon atom is created near the surface of a grain, it can recoil into the pore between grains; such radon atoms do not attach or bind to the matrix that contains the immediate precursor, radium. The amount of radon that reaches the pores is described by the emanation fraction. For typical soils or bedrock, the emanation fraction can range from 5% to 50% (see the review in Nazaroff 1992).

In most situations, the pore between grains of material contains a mixture of air and water. Often, a recoil radon atom will come to rest in the water and remain there (Tanner 1980). In addition to this direct process, a gas is partitioned between the air and water in the pore. This partitioning is described by Henry's law in terms of the Oswald coefficient, K:

$$K = \frac{C_w}{C_a} \tag{1.1}$$

where C_w and C_a are the radon concentrations by volume (Bq m^{-3}) in the water and air, respectively. The Oswald coefficient varies inversely with temperature. At 10 °C, $K_{Rn} = 0.3$; it increases to about 0.5 near 0 °C (Lewis and others 1987). If the soil or bedrock is completely saturated with water, all the available radon will be dissolved in the water.

Migration of radon in soil gas is controlled by two processes: molecular diffusion and advective flow. Diffusion is the process whereby molecules migrate toward regions with lower concentrations. Radon concentrations in soil gas are typically 40,000 Bq m^{-3} and concentrations 10 to 100 times this value are not uncommon. The main reason for this is that the radon atoms are confined within a small volume defined by the pore space between the soil grains. Thus, radon will preferentially diffuse toward regions that have lower concentrations, such as caves, tunnels, buildings, and the atmosphere.

Advective flow is controlled by pressure differences. Air will flow toward locations with lower pressure, and changes in atmospheric pressure can force air into or out of the ground. Very often, the air inside a building is warmer than air in the soil that is in contact with the building. This temperature difference causes a pressure gradient that draws air containing high concentrations of radon into the structure. Wind—as well as airflow from a fan, furnace, or fireplace—can also reduce pressures inside a building, compared with the pressures in the soil adjacent to the building foundation. These processes constitute the primary reason that radon enters and may be present in buildings at higher concentrations than in ambient air.

The water supply can also contribute to indoor radon. When water leaves a faucet, dissolved gases are released. This process is increased by mechanical sprays during a shower or by the heating and agitation that occur during laundering, washing, and cooking. The increase in the indoor radon concentration due to radon release from indoor water use is described by the transfer coefficient:

$$T = \frac{\overline{\Delta C_a}}{\overline{C_W}} \tag{1.2}$$

where $(\overline{\Delta C_a})$ is the average increase of the indoor radon concentration that results from using water having an average radon concentration of $\overline{C_W}$. The various sources of radon and the resulting radiation exposure pathways are shown in figure 1.2.

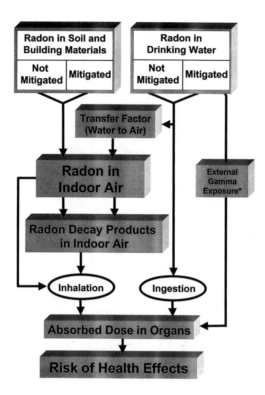

FIGURE 1.2   Sources of radon and related radiation exposure pathways.
*Gamma exposure from radon collected during some mitigation procedures (see Appendix E).

EXPOSURE TO INDOOR RADON

The first four descendants of radon—^{218}Po, ^{214}Pb, ^{214}Bi, and ^{214}Po—are also radioactive and are collectively referred to as *radon decay products*. They are all metals and have half-lives ranging from a fraction of a second to 27 min (see figure 1.1). Indoors, some of these decay products come into contact with surfaces and are removed from the air by a process called plate-out. The rest of the decay products remain suspended in air as free atoms (unattached) or combined with other aerosols (attached). Although it is possible to measure the concentration of each radon decay product suspended in air, they are generally grouped. In addition, the concentration is not presented in terms of activity per unit volume (becquerel per cubic meter), but rather in terms of the total energy that would be released by alpha particles when all the short-lived atoms decayed completely. This quantity is called potential alpha energy (PAE), and the concentration in air

(PAEC), is measured in units of energy per unit volume of air (joules per cubic meter (J m^{-3}).

The development of a PAEC from indoor radon concentration depends on air movement and aerosol conditions within a room. PAEC can depend on whether the radon entered a room from soil or from water during bathing. The relationship between indoor radon concentration and PAEC is expressed in terms of the equilibrium ratio (ER). For a room without any depletion of radon or plate-out of decay products, ER = 1.0. In domestic environments, ER ranges from 0.3 to 0.7 with a nominal value of 0.4 (Hopke and others 1995a).

The alpha-particle dose to lung tissues depends on PAEC and on the time that a person spends in a given location. A combination of PAEC and time is a measure of exposure expressed in joule-seconds per cubic meter (J· s m^{-3}).

ABSORBED DOSE FROM INDOOR RADON

A person in a room will inhale radon decay products that are suspended in air. Some activity can deposit and accumulate in the respiratory airways, depending on breathing patterns and the aerodynamic size of the particles with which the decay products are associated. Because of the short half-lives, the radon decay products that are deposited in the lung will almost certainly decay completely in the lung. The radiations emitted within the lung during these decays can deposit energy in the body. However, this radioactivity is very near the lung epithelium, so alpha particles in particular can transfer copious amounts of energy to vulnerable cells. That is why radon decay products are characterized in terms of PAEC.

Radon gas itself is also inhaled. Most of it is exhaled immediately and therefore does not accumulate in the respiratory system, as do radon decay products. Because the radon does not get close to radiosensitive cells, the absorbed dose from alpha particles is small. However, some of the radon that reaches the interior region of the lung is transferred to blood and dispersed throughout the body. Radon and the decay products formed inside the body can deliver a radiation dose to tissues and organs.

On some occasions, water is consumed immediately after leaving the faucet before its radon is released into the air. This water goes directly to the stomach. Before the ingested water leaves the stomach, some of the dissolved radon can diffuse into and through the stomach wall. During that process, the radon passes next to stem or progenitor cells that are radiosensitive. These cells can receive a radiation dose from alpha particles emitted by radon and decay products that are created in the stomach wall. After passing through the wall, radon and decay products are absorbed in blood and transported throughout the body, where they can deliver a dose to other organs.

Ingested water eventually passes through the stomach into the small intestine, where the remaining radon and decay products are released from the water and transferred to blood. They then circulate within the body; most are released from the blood into the lung and exhaled, but some remain in the blood and accumulate in organs and tissues, which receive an absorbed dose from alpha, beta, and gamma radiation.

HEALTH RISKS POSED BY INDOOR RADON

There is a direct implication between high doses of radiation and health effects in humans. For example, excess cancers have been observed in a cohort of survivors of the atomic-bomb blasts in Japan (National Research Council 1990a). A relationship between lung cancer and inhalation of radon decay products has been demonstrated in underground miners (Lubin and Boice 1997). Recent epidemiologic evidence suggests that inhalation of radon decay products in domestic environments could also be a cause of lung cancer (National Research Council 1999; Lubin and others 1995). Although the studies do not specifically identify health effects at low doses, there is compelling circumstantial evidence that they occur.

Under ambient conditions of low dose and low dose rate, any health effects associated with exposure to radon in air or water can be expected to occur from the passage of single alpha particles through individual cells. Any given cell is hit only once or not at all. An increase in exposure increases the number of cells that are hit, but it will not affect the primary damage experienced by each cell. Therefore, the initial events depend linearly on exposure or dose.

Exposed cells experience local damage in the form of DNA breaks and the products of reactive oxygen. The damage is metabolized by cellular-repair systems, and some fraction of it results in permanent genetic changes. Those changes can lead to the development of cancers; a cancer usually originates in a single transformed cell.

Risk projection models have been developed to predict the risk in situations where direct evidence is not available (National Research Council 1999; 1990a). The nature of the exposure to indoor radon, the kinds of DNA damage inflicted by alpha particles, and the extent of repair are consistent with the absence of a threshold for cancer induction. The preferred model is a straight line that reaches zero risk only when the dose or exposure is zero; it is referred to as the linear no-threshold (LNT) model.

LEGISLATION AND REGULATIONS REGARDING INDOOR RADON

In 1988, Congress passed the Indoor Radon Abatement Act. Its stated goal was to reduce indoor radon concentrations to outdoor levels. The Environmental Protection Agency (EPA) was authorized to implement policies described in the law. In 1987 and again in 1992, EPA published *A Citizen's Guide To Radon* (EPA 1992a). The document summarized the risks associated with inhalation of radon decay products in residential environments. It recommended that people measure indoor radon and consider taking action if the annual average concentration in their living areas exceeds 148 Bq m^{-3}. EPA also developed programs in support of its recommendations for mitigation (Page 1993): public-information programs, a National Residential Radon Survey, Regional Radon Training Centers, the Radon Contractor Proficiency Program, the Radon Measurement Proficiency Program, Radon Reduction in New Construction, and support for the development of indoor-radon programs in individual states. As a result of those efforts, about 11 million of the approximate 100 million single family dwellings in the United States have been tested and about 300,000 (0.3%) mitigated in an effort to reduce indoor radon concentrations (CRCPD 1994). In addition, EPA estimates about 1.2 million new homes have been built with radon-resistant construction methods (A. Schmidt, personal communication), although the success of these methods is unknown.

It was recognized that water might also make a substantial contribution to and in some circumstances be the primary source of health risks associated with radon. In 1986, a revision to the Safe Drinking Water Act specifically required EPA to set a standard for ^{222}Rn in drinking water (US Congress 1986). After litigation and a consent decree, EPA developed a criteria document that summarized the health effects of radon and its prevalence in drinking water (EPA 1991a). On the basis of the document and considerations of uncertainties in the analytic procedures for testing for radon in drinking water, a regulation was proposed in 1991 that established a maximum contaminant level (MCL) of 11,000 Bq m^{-3} (EPA 1991b). That MCL corresponded to an lifetime individual health risk of 10^{-4} posed largely by an increase in radon in indoor air.

During the period permitted for public response after the announcement of the proposed regulation, some groups supported reducing the MCL below 11,000 Bq m^{-3} because there is no known threshold for radiation-induced carcinogenesis. Others suggested raising the MCL because the increment in indoor-air radon from water radon at 11,000 Bq m^{-3} would be about 2% of the annual average residential radon concentration. There was also concern regarding the dosimetry model used to estimate the risk of stomach cancer associated with radon ingestion (Harley and Robbins 1994). As a result of those concerns, Congress intervened in 1992 and directed the administrator of EPA to prepare a multimedia risk assessment and cost estimates for compliance with regulations regarding radon in drinking water. The reanalysis resulted in EPA's revising its risk assessment for the ingestion of water containing radon. As a result, the ingestion risk and the inhalation risks (per unit of radon in drinking water) were estimated to be about equal (EPA 1994b). This document was reviewed by the Science Advisory Board (SAB) of EPA.

There was continuing concern about the estimates of stomach cancer resulting from radon ingestion. In addition, the SAB committee questioned the prudence of regulating a small increase in indoor radon from water without considering the larger reductions in risk that might be obtained by reducing radon concentrations originating from soil gas (EPA-SAB 1993a).

The Safe Drinking Water Act was amended again in 1996 (US Congress 1996). The proposed national primary drinking-water regulation for radon was withdrawn. Before proposing a new regulation for radon in water, EPA was instructed to ask the National Academy of Sciences to prepare a risk assessment for radon in drinking water on the basis of the best science available. The assessment was to consider each of the pathways associated with exposure to radon from drinking water at concentrations and conditions likely to be experienced in residential environments. The Academy was also asked to prepare an assessment of health-risk reductions that have been realized from various methods used to reduce radon concentrations in indoor air to provide a basis for considering alternative or multi-media mitigation schemes as opposed to mitigation of water alone.

CHARGE TO THE COMMITTEE

The Committee on the Risk Assessment of Exposure to Drinking Water in the National Research Council's Board on Radiation Effects Research began deliberations in July 1997. The specific tasks assigned to the committee were:

- To examine the development of radon risk assessments for both inhalation of air and ingestion of water.
- To modify an existing risk model if it were deemed appropriate or develop a new one if necessary.
- To review the scientific data and technical methods used to arrive at risk coefficients for exposure to radon in water.
- To assess potential health-risk reductions associated with various measures to reduce radon concentrations in indoor air.

The final report was to include:

- Estimates of lung, stomach, and other potential cancer risks per unit concentration of radon in water.
- Assessment of whether health effects of radon in drinking water could be estimated for various sub-populations at risk, such as infants, children, pregnant women, smokers, elderly persons, and seriously ill persons.
- Examination of evidence for teratogenic and reproductive effects in men and women due to radon in water.
- Estimates of the transfer coefficient that relates radon in water to radon in indoor air.
- Population-weighted estimates of radon concentrations in ambient air.
- Estimates of increases in health risks that could result from methods used to comply with regulations for radon in drinking water.

• Discussion of health-risk reductions obtained by encouraging people to reduce radon concentrations in indoor air with methods already developed and comparison of them with the risk reductions associated with mitigation of radon in water.

COMPOSITION OF THE REPORT

Chapter 2 presents baseline data regarding concentrations of radon in water and indoor air. It includes a discussion of radon concentrations measured in outdoor air throughout the United States and an estimate of a national annual average concentration of ambient radon.

Chapter 3 describes the transfer coefficient that expresses the increase in indoor airborne radon in reference to the concentration of radon in water. It includes a survey of measurements and theoretical considerations.

Chapter 4 discusses the dosimetry of ingested radon. It describes patterns of consumption of water directly from the tap or faucet. The calculations make extensive use of physiologically-based pharmacokinetic (PBPK) models that have been developed for dosimetry of internal radioactivity. The chapter includes computations of equivalent dose and risk to individual tissues and organs. A special model was developed to estimate the concentration of radon and the alpha-particle radiation dose produced by decay of radon and its decay products occurring next to sensitive cells in the stomach wall.

Chapter 5 discusses the risk associated with inhalation of radon and radon decay products. It includes a summary of the methods used to form risk-projection models that were developed by the National Research Council's committees on Biological Effects of Ionizing Radiation (BEIR).

Chapter 6 discusses the basic mechanisms that are believed to be responsible for radiation-induced carcinogenesis.

Chapter 7 presents an analysis of the uncertainty and precision associated with the risk estimates obtained in the previous chapters.

Chapter 8 discusses the methods and efficiencies of radon mitigation in both indoor air and water. It includes an examination of techniques for reducing radon concentrations in existing buildings and procedures for reducing radon in new construction.

Chapter 9 analyzes the concepts associated with a multimedia approach to risk reduction. Several scenarios illustrate various ways to evaluate gains in risk reduction by using an alternative AMCL for water with other indirect approaches that encourage or even enforce mitigation in indoor air.

The committee's research recommendations are summarized in chapter 10.

A glossary and six appendixes present specific details and methods that were incorporated in the various chapters.

2

Baseline Information on Indoor Radon and Radon in Water in the United States

Several databases provide a national picture of indoor radon and radon in water for the United States. We provide these data here as context for the discussions in later chapters on ambient radon, transfer factors, uncertainty, mitigation, and a multimedia approach to risk reduction. Figure 2.1 is a geologic-physiographic map of the United States that will serve as a general reference for areas of the country that are important as sources of radon (Schumann and others 1994); it is derived from standard geologic and physiographic maps.

INDOOR RADON

The concept of radon potential can be used as a basis for estimating indoor radon concentrations. Although it is not possible to accurately predict radon concentrations in individual houses because of the highly variable nature of factors that control radon entry and concentrations in a specific house, one can estimate the distribution of indoor radon concentrations on a regional basis. Several approaches have been taken to develop indoor-radon potential maps of the United States, and succeeding studies have built on previous ones; the most recent maps of predicted indoor radon encompass a statistical analysis of variables that account for the greatest variation in indoor radon: geology, climate, and house structure.

Figure 2.2 shows the geologic-radon potential map of the United States developed by the US Geological Survey (Gundersen and others 1992) on the basis of geology, indoor radon measurements, the aerial radiometric data collected by the National Uranium Resource Evaluation (summarized in Duval and

32

FIGURE 2.1   Geologic-physiographic map of the United States (courtesy of USGS).

others 1989), soil permeability, and foundation housing characteristics. It is a map of the land potential, not a map of exposure or risk. It was compiled from individual state geologic-radon potential maps (Gundersen and others 1993) that served as the basis of the Environmental Protection Agency (EPA) map of radon zones that has been incorporated into one of the national building codes (EPA 1993).

Figure 2.3 shows the most recent and most comprehensive map of indoor-radon potential and represents a prediction of the geometric mean of annual exposure to indoor radon. The elements used in the map include the radium content of the surficial soil derived from the aerial radiometric data collected by the National Uranium Resource Evaluation (summarized in Duval and others 1989), information on the geologic province that comprises most of the county (from the US Geological Survey), soil characteristics, the fraction of homes with basements and with living-area basements, and radon-concentration surveys conducted nationally and in each state from EPA and other sources. Those elements are used in a Bayesian mixed-effects regression model to provide predictions of the geometric mean indoor radon concentration by county. Additional details of the model are given in Price (1997). The predicted county means have standard errors of 15-30% for typical counties; the uncertainty in a given county depends on the number of radon measurements in the county and the level of detail in the geologic information.

34

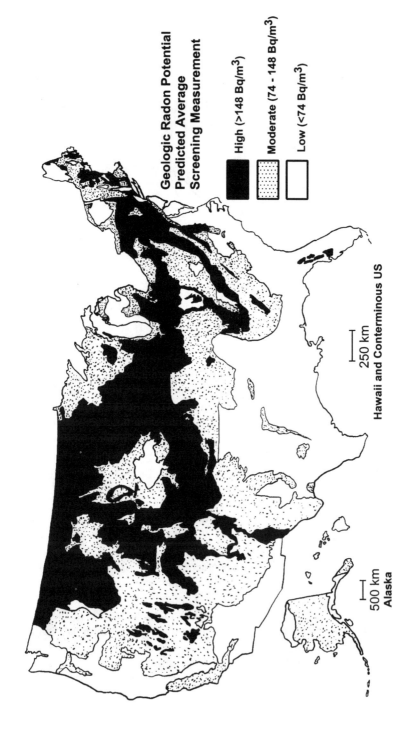

**Geologic Radon Potential
Predicted Average
Screening Measurement**

■ High (>148 Bq/m³)

▨ Moderate (74 - 148 Bq/m³)

□ Low (<74 Bq/m³)

250 km

Hawaii and Conterminous US

500 km

Alaska

FIGURE 2.2   Geologic-radon potential map of the United States (Courtesy of USGS).

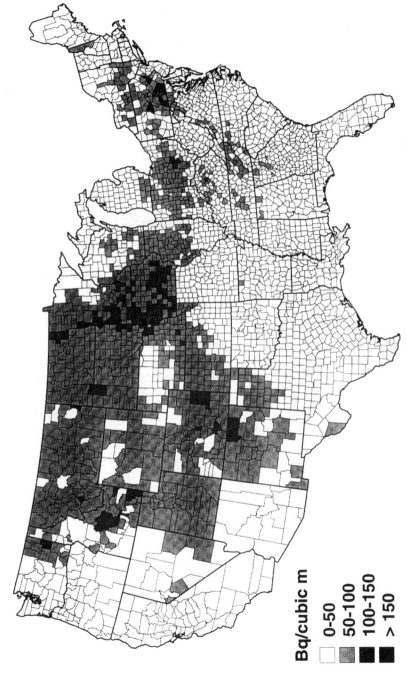

FIGURE 2.3 Indoor-radon potential [predicted geometric mean air concentration in living area, Bq/m³].

Bq/cubic m

0-50	
50-100	
100-150	
> 150	

From figures 2.2 and 2.3, it is obvious that the Appalachian Mountains, Rocky Mountains, Colorado Plateau, and northern glaciated states (states north of the limit of glaciation) tend to have the highest radon potential and indoor radon. The principal geologic sources of radon in the United States are:

- Uranium-bearing metamorphosed rocks, volcanics, and granite intrusive rocks that can be highly deformed or sheared (shear zones in these rocks cause the largest indoor-radon problems in the United States), found predominantly in the Appalachian Mountains, Rocky Mountains, and Basin and Range;
- Glacial deposits derived from uranium-bearing rocks and sediments found in the northern tier of states above the limit of glaciation;
- Marine black shales found in the Appalachian Plateau and Great Plains and to a smaller extent in the Coastal Plain, Colorado Plateau, and Basin and Range;
- High-iron soils derived from carbonate, especially in karstic terrain found in the Appalachian Plateau, Appalachian Mountains, and Coastal Plain; and
- Uranium-bearing fluvial, deltaic, marine, and lacustrine deposits and phosphatic deposits found in the Colorado Plateau, Rocky Mountains, Great Plains, Coastal Plain, Basin and Range, and Appalachian Plateau.

RADON IN GROUNDWATER AND PUBLIC WATER SUPPLIES

In the 1980s, a number of national studies of radon and other radionuclides in public water supplies and groundwater in the United States were published (see (Longtin 1988; Michel and Jordana 1987; Hess and others 1985; Horton 1983). These studies examined geographic distribution, the controls of hydrogeology, and differences among private well, small public, and large public water supplies. The most common conclusions of the studies suggest that the highest radon concentrations in groundwater and public water supplies generally occur in portions of the Appalachian Mountains, Rocky Mountains, and Basin and Range. Private well sources and small public water supplies tend to be higher in radon than large public water supplies. Private well sources and small water supplies tend to be in aquifers with low capacity. When these types of aquifers are uranium bearing granite, metamorphic rocks, or fault zones (as found in the mountain states), the radon concentration in the water tends to be high. Large public water supplies tend to use high-capacity sand and gravel aquifers, which generally comprise low-uranium rocks and sediments and tend to be lower in radon.

The study of Hess and others (1985) examined 9,000 measurements of radon in water from national and state surveys. Data were compiled for all but 10 states. Public water supplies originating in surface water tended to have radon concentrations less than 4,000 Bq m^{-3}. Private water supplies were higher in radon than public water supplies by factors of 3 to 20. States with the highest radon in private well water were Rhode Island, Florida, Maine, South Dakota, Montana, and

Georgia. The New England states overall had the highest radon concentrations in water from all sources; state geometric means ranged from 18,500 Bq m^{-3} in Massachusetts to 88,800 Bq m^{-3} in Rhode Island. A population-weighted geometric mean for the United States of 6,900 Bq m^{-3} was reported.

Two major national databases collected by EPA exist for radioactivity in public water supplies. Beginning in November 1980, EPA systematically sampled the 48 contiguous states, focusing on water supplies that served more than 1,000 people (Horton 1983). Radon samples were analyzed with liquid scintillation-counting methods, and samples were targeted to be from as close to the ground-water source as possible and to exclude surface waters. The more than 2,500 public water supplies that were sampled represented 45% of the water consumed by US groundwater consumers. High radon concentrations were found in the waters of the New England states, North Carolina and South Carolina, Georgia, Virginia, Arizona, Colorado, Nevada, Montana, and Wyoming. Individual sample measurements ranged from 0 to over 500,000 Bq m^{-3}, the average was 12,600 Bq m^{-3} and the geometric mean was 3,700 Bq m^{-3}.

From 1984 to 1986, EPA conducted the National Inorganics and Radionu-clides Survey on the basis of 990 randomly distributed samples from the inventory of public water systems in the Federal Reporting Data System (Longtin 1988). The random sample was stratified into four general categories that represented the population served by the system and represented finished water in the distribution system, generally sampled at the tap. Radon was measured with liquid scintillation-counting methods. Longtin (1988) calculated a population-weighted average radon concentration of 9,200 Bq m^{-3} but did not calculate unweighted statistics. Our committee examined the unweighted data; of the 990 records, 275 had censored observations of less than 3,700 Bq m^{-3}. Values ranged from below the detection level to 949,000 Bq m^{-3}. The distribution of the concentrations was assumed to be log normal and statistics were estimated with the method of maximal likelihood, using SAS and LIFEREG, which accounts for censored data. The geometric mean radon concentration was estimated at 7,500 Bq m^{-3}, the average 20,000 Bq m^{-3} and the geometric standard deviation 4.06.

A comparison of the two data sets with the data of Hess and colleagues (1985) is shown in figure 2.4. The distributions appear similar in most respects. The 9,000 measurements of Hess and others included the 2,700 measurements of Horton (1983) and some state studies but did not include the Longtin (1988) data. The Hess data have higher percentages of readings in the highest concentration categories than either of the other two data sets. The Horton data have the highest percentage of radon measurements less than 18,500 Bq m^{-3}.

AMBIENT RADON

Ambient radon concentration is the concentration of radon in the atmosphere. The outdoor concentration of radon varies with distance and height from

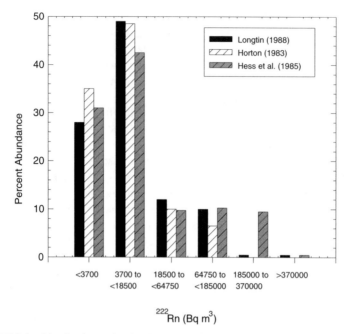

FIGURE 2.4   Distributions of radon in water measurements in several studies across the United States.

its principal source in the ground (rocks and soil) and distance from other sources that can locally or regionally affect ambient radon, such as bodies of water, mine or mill tailings, vegetation, and fossil-fuel combustion. The decrease in radon with height from the source is not simply tied to ground exhalation, nor is the variance a simple mathematical function. A number of studies have documented the decrease in ambient radon with increasing height above the ground and concluded that it is due predominantly to dilution by atmospheric mixing and turbulence (Gogolak and Beck 1980; Druilhet and others 1980; Bakulin and others 1970; Pearson and Jones 1966; Servant 1966; Moses and Pearson 1965; Pearson and Jones 1965). The ambient radon concentration can decrease by more than half in the first 10 m, but many studies show decreases of only one-tenth to one-third in the first 10 m. Concentrations of outdoor radon also change daily and seasonally in response to temperature, changes in atmospheric pressure, and precipitation.

Gesell (1983), Blanchard (1989), and Harley (1990) reviewed available studies of outdoor radon from around the world and observed consistent diurnal and seasonal trends. Generally, the diurnal pattern of outdoor radon concentration includes early morning and evening maxima related to cooling and air stability. Minimum concentrations typically occur in the afternoon because of warming, evaporation and transpiration from soil, and mixing of air. Maximum concentrations occur after midnight and in the early morning hours because of inversion, cooling, and the increased stability of air masses. The marked diurnal pattern is illustrated in figure 2.5, which shows 7 y of hourly data from a site in suburban northern New Jersey (N. Harley, personal communication). As in diurnal patterns around the world, the radon concentration overnight was much greater than that during the day. The ratio of maximums to minimums generally ranges from 1.5 to 4.

Seasonally in the United States, maximum outdoor radon concentrations often occur in the summer to early winter and minimum concentrations in the late winter to spring in reaction to meteorologic changes and moisture conditions in the ground. The seasonal pattern varies somewhat in different parts of the world because of variations in seasonal wet and dry periods. Some moisture greatly increases radon emanation (Tanner 1980) whereas too much moisture or saturation of the soil greatly decreases radon transport to the atmosphere. Large barometric-pressure changes and precipitation events yield short-lived but large variations in ambient radon and soil radon on a particular day or seasonally (Schumann and others 1992; Clements and Wilkening 1974). The influence of barometric pressure is illustrated in figure 2.6, which shows outdoor-radon data from Fort Collins, Colorado (Borak and Baynes 1999). A change in barometric pressure changes the pressure gradient between the atmosphere and soil. The soil response depends on the magnitude and duration of the change; a dramatic increase in barometric pressure suppresses radon transport to the atmosphere, and a decrease

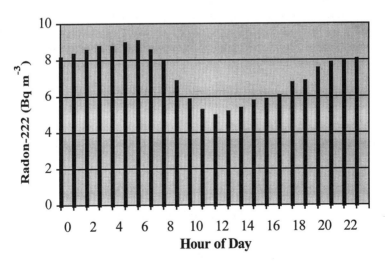

FIGURE 2.5   Diurnal variation of ambient radon at a site in northern New Jersey averaged over 7 y.

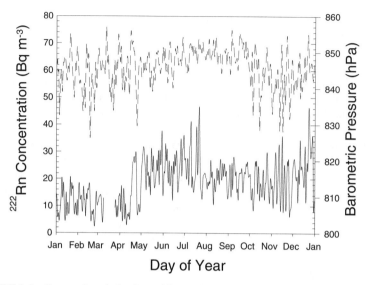

FIGURE 2.6   Seasonal variation in ambient radon and barometric pressure during 1994
from a site at Fort Collins, Colorado. (Radon concentration is lower curve.)

in pressure enhances radon transport from the soil. The radon response often lags
slightly behind the barometric-pressure change and is diminished if the pressure
change is relatively small or gradual, which allows equilibration.

   The ratio of maximums to minimums for seasonal variation of outdoor radon
generally ranges from 2 to 5 and is larger in summer than in winter. Ambient
radon concentrations differ geographically because of differences in ground concentrations of radon related to geology, soil texture, moisture, atmospheric dilution by adjacent water bodies, and climatic and meteorologic sources. Overviews
of ambient radon concentrations around the world include those by Gesell (1983),
NCRP (1988), UNSCEAR (1988), and Harley (1990). Those studies report
averages above continental land masses generally in the range of 4-75 Bq m^{-3}
and averages above water bodies or islands generally less than 2 Bq m^{-3}. Wilkening and Clements (1975) estimated that the ocean contributes only 2% of
atmospheric radon.

Studies of Ambient Radon

   For the purposes of this report, we have compiled and examined most of the
outdoor-radon studies conducted in the United States during the last 15 y. There
has been only one national study in the United States, but several ambient-radon
studies have been carried out on the state or regional scale or at a single site over

time. In general, the studies examined in this section reported readings taken at about 1-2 m above the ground with various methods, most commonly the use of alpha-track detectors, continuous or semicontinuous radon monitors, and electret ion chambers. The accuracy and precision of the individual methods has been examined in numerous studies and recently reviewed by Fortmann (1994), Lucas (1957) and Busigin and others (1979). Studies on the quality of data in the range of 1-40 Bq m^{-3} are rare, although measurement detection limits for the different devices range from 1 to 18 Bq m^{-3} (Blanchard 1989). Measurement errors reported in the studies that the committee compiled generally range from 8 to 20% but can be substantially higher when very low concentrations were measured. Quality control and duplicate measurements were used in all the studies. Only a few studies measured radon progeny and calculated equilibrium factors and outdoor dose rates or dose (Wasiolek and others 1996; Wasiolek and Schery 1993).

National Studies

In the late 1980s, EPA conducted a national survey of ambient radon across the United States (Hopper and others 1991) to confirm previously reported concentrations and in response to section 302 of the Indoor Radon Abatement Act, which stated that "the national long-term goal of the United States with respect to radon levels in buildings is that the air within buildings in the United States should be as free of radon as the ambient air outside of buildings." Section 303 also required EPA to include information regarding outdoor ambient radon concentrations around the country in the updated *Citizens Guide to Radon*. From 1989 to 1991, measurements were made quarterly in 50 cities, one in each state, across the country. The sites chosen coincide with 50 EPA's Environmental Radiation Ambient Monitoring System stations that were established in 1973 and most coincide with the capital cities of the states. The ambient radon concentrations measured at the sites are shown in table 2.1. Summary statistics and the frequency distribution of the seasonal averages are shown in figure 2.7. Measurements were made at each station with three electret ion chambers placed in ventilated shelters 1 m above ground. In each shelter, there were also three thermoluminescent dosimeters to measure gamma radiation to provide the needed gamma correction of the electret measurements. Every 90 d, the devices were exchanged for new ones, and the old ones were measured at EPA's Las Vegas facility. Three devices were used to assess precision and allow for backup in case a device failed, and readings were taken quarterly to examine seasonal variation. (EPA 1992d; Hopper and others 1991). Measurements were reported in pCi/L, but for this report we have converted the measurements to Bq m^{-3} [note: 1 m^3 = 1,000 L]. The limit of detection of the devices was determined to be 2 Bq m^{-3}. During the first quarter, several stations were started several weeks late and problems with the setting up of the stations and the measuring protocol were found in several states (these data were not included). Corrections of the proce-

TABLE 2.1   Seasonal Ambient Radon, Bq m^{-3} for the United States
(Arithmetic Average of Three Detectors at Each Site)

Site	Quarter 1 Sum. 1989	Quarter 2 Fall 1989	Quarter 3 Winter 1989	Quarter 4 Spring 1990	Quarter 5 Sum. 1990
AL	13.3	16.4	12.5	12.0	26.5
AK	12.7	6.8	10.6	8.5	11.1
AR	11.7	21.6	19.5	11.8	15.2
AZ	8.8	16.2	28.5	20.0	13.6
CA	11.8	16.8	20.5	11.3	11.7
CO	12.3	27.5	17.3	8.0	10.5
CT	13.4	25.9	13.0	ND	15.4
DE	17.5	13.9	13.6	10.5	13.9
FL	25.9	14.7	14.1	9.6	11.0
GA	14.8	28.7	19.0	13.9	13.0
HI	9.5	6.8	7.0	5.9	8.9
IA	24.8	14.6	25.9	14.1	17.3
ID	13.3	16.3	27.4	9.1	9.7
IL	22.7	21.8	24.8	16.5	20.5
IN	19.4	15.7	15.0	19.6	15.7
KY	19.5	16.4	20.0	11.6	18.6
KS	27.3	19.2	23.4	15.8	19.6
LA	17.4	13.8	8.3	5.7	9.1
MA	22.9	17.0	15.2	8.0	14.4
MD	ND	19.7	22.6	12.7	16.7
ME	19.2	18.1	20.1	13.2	15.4
MI	16.3	12.1	15.3	10.5	14.9
MN	18.5	14.1	17.6	8.8	10.6
MO	17.4	28.0	25.3	17.0	16.7
MS	15.5	16.0	11.0	11.7	15.7
MT	12.8	18.9	17.0	15.3	19.4
NC	18.7	14.6	15.9	5.3	12.2
ND	19.4	27.5	19.1	15.5	14.7
NE	20.0	22.0	21.2	14.6	18.7
NH	24.4	15.4	14.1	10.1	15.0
NJ	15.3	18.5	17.4	12.5	15.7
NM	7.8	6.2	10.2	2.5	4.4
NV	5.1	10.0	12.6	5.7	6.3
NY	12.7	11.8	15.0	9.3	10.4
OH	16.9	14.1	19.7	10.1	12.2
OK	10.4	13.4	13.6	10.4	23.8
OR	ND	9.0	17.5	10.2	11.6
PA	20.9	21.5	24.2	10.0	16.5
RI	ND	5.1	16.3	10.6	9.7
SC	15.4	35.9	17.5	12.2	16.7
SD	17.9	21.2	21.7	16.0	18.5
TN	17.3	18.0	19.8	11.1	20.2
TX	17.9	17.0	21.1	7.0	20.7
UT	7.8	15.0	15.2	5.9	9.3
VA	14.6	20.0	16.2	13.9	18.6
VT	18.9	14.1	14.3	13.8	15.0
WA	15.0	24.9	16.0	14.3	20.8
WI	13.4	22.4	15.0	11.2	13.0
WV	22.0	11.6	ND	15.0	20.5
WY	7.4	16.7	ND	8.8	10.5
Avg.	16.1	17.3	17.5	11.4	14.8

Quarter 6 Fall 1990	Quarter 7 Winter 1990	Quarter 8 Spring 1991	Quarter 9 Sum. 1991	Avg. All Quarters
13.7	13.4	7.8	13.3	14.3
11.6	ND	ND	ND	10.2
20.5	15.5	16.0	15.7	16.4
18.3	18.0	14.6	12.5	16.7
18.5	13.4	10.5	12.5	14.1
16.9	13.2	5.8	3.6	12.8
15.7	13.4	7.5	12.6	14.6
13.7	15.0	7.9	15.9	13.6
9.1	9.0	11.2	18.5	13.7
11.3	14.8	10.1	15.4	15.7
8.5	7.3	6.3	10.2	7.8
20.4	18.3	13.1	20.8	18.8
17.4	23.3	5.9	13.9	15.1
17.5	10.1	11.8	20.5	18.5
13.8	11.5	10.2	15.9	15.2
18.9	14.4	12.7	24.5	17.4
24.7	22.0	15.3	19.2	20.7
6.8	9.7	8.0	9.7	9.8
15.7	13.7	10.7	14.8	14.7
15.2	17.0	11.1	16.8	16.5
17.3	10.5	16.2	17.4	16.4
16.2	11.8	11.5	14.8	13.7
15.0	14.7	10.5	11.7	13.5
20.6	16.2	11.0	14.7	18.5
17.6	9.9	7.5	14.8	13.3
20.2	17.6	12.5	20.1	17.1
11.8	9.3	8.3	ND	12.0
25.2	22.4	18.0	14.1	19.5
22.8	21.7	15.8	24.4	20.1
16.5	11.5	11.8	14.6	14.8
19.6	11.3	17.4	11.0	15.4
6.7	1.7	4.4	3.9	5.3
12.4	8.5	4.2	ND	8.1
11.5	8.9	12.3	ND	11.5
14.2	9.6	11.6	16.8	13.9
13.3	10.9	25.2	10.6	14.6
13.1	11.2	9.6	11.2	11.7
16.4	11.6	15.5	19.9	17.4
9.0	8.3	15.2	11.5	10.7
13.7	11.8	10.0	15.2	16.5
24.7	23.8	30.7	21.0	21.7
21.2	11.2	10.5	21.1	16.7
13.2	11.7	7.0	8.9	13.8
11.5	14.9	6.4	5.6	10.2
29.5	12.5	13.7	16.2	17.2
14.4	11.0	5.3	14.1	13.4
18.9	14.8	12.8	17.9	17.3
14.6	10.4	15.0	18.6	14.8
19.2	18.4	13.3	27.3	18.4
ND	10.4	10.4	5.2	9.9
16.1	13.3	11.6	15.0	14.8

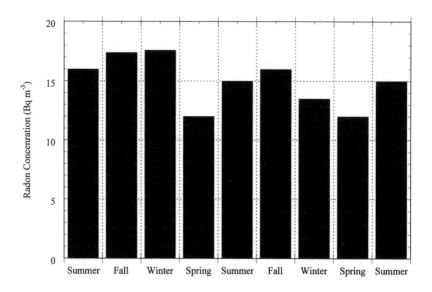

FIGURE 2.7   Average ambient radon concentrations at 50 sites in the United States.

dure were made by the second quarter. The authors (Hopper and others 1991) emphasize that the study does not statistically represent the distribution of ambient radon concentrations in the United States but indicates that estimates of annual average ambient radon concentrations and the associated error estimates can be derived for each site. The authors used only quarters 2-5 for their original report for the sake of timeliness, but provided our committee with the entire data set for this report (R. Hopper, private communication). The updated *Citizens Guide To Radon* (EPA 1992c) reported an average outdoor concentration of 14.8 Bq m^{-3} on the basis of the survey.

As can be seen from table 2.1, ambient radon concentrations above the average (all quarters, all sites, 14.8 Bq m^{-3}) tend to occur in the Appalachian Mountains, the northern Midwest, and the northern western states. Sites in the southern and western coastal states, the Great Lakes states, and several of the central and southwestern states tend to be at or below the average. These trends probably reflect the geology or other sources at the sites and the proximity to large water bodies. The bar graph in figure 2.7 illustrates the average of all data from each site by season, showing the spring minima and fall maxima.

State Studies

Statewide or regional studies have been conducted in California (Liu and others 1991), Nevada (Price and others 1994), Minnesota and Iowa (Steck and others 1999), and Maine (Hess and others 1982). In the Nevada study, Price and others (1994) used the same method as Hopper and others (1991) and measured 50 sites across the state during a 30-d period in the summer of 1992. Sites were chosen to reflect different rock types and represent the principal population centers in Nevada. About half the sites were in residential areas, and the rest were in remote areas near rock outcrops. Results indicate that radon in soil gas corresponds well with the geology, outdoor radon concentration and indoor radon. Measurements across the state ranged from 2.6 to 52 Bq m^{-3}; the geometric mean was 13.1 Bq m^{-3}. The range and values of concentrations were generally very similar to what Hopper and others found for the United States.

As part of a statewide radon study, indoor radon was measured at 300 sites throughout California (Liu and others 1991). At 68 of those sites, outdoor radon was also measured by using alpha-track detectors in cups suspended 1-2 m above the ground and exposed for a year starting in April 1988. Indoor radon was found to correlate well with broad geologically defined areas of the state. The geometric mean outdoor radon concentration was 15.54 Bq m^{-3}, and the range was 0.3 to 55.5 Bq m^{-3}.

Steck and others (1999) measured annual average atmospheric radon concentrations at 111 locations across Iowa beginning in 1993 and ending in the spring of 1997. They also measured ambient radon at 64 selected sites in western and northern Minnesota during 1995-1996. Comparisons were made with indoor radon; at some sites seasonal variations and variation with height were tested. Large-volume alpha-track detectors were enclosed in protective housings and placed 1.5-2 m above the ground for a year at each site. In Minnesota, concurrent annual average indoor radon measurements were also made with the same type of device. In Iowa, the researchers found that elevated outdoor radon concentrations twice the annual average reported by Hopper and others persisted over long periods and covered wide areas of the state. In both states, some outdoor concentrations were the same as or higher than the national average indoor radon concentration (46 Bq m^{-3}). In general, outdoor radon concentrations were distributed in a geographically similar pattern to indoor measurements.

Etched-track detectors were used to measure outdoor radon and multi-room indoor radon at 100 sites in Maine from October 1980 to May 1981 to determine integrated average radon concentrations during the heating season (Hess and others 1982). The outdoor cups were in open sheltered areas on porches, garages, and sides of homes approximately one meter above the ground. Over half the houses were in geologic regions where high concentrations of radon in water were previously found. The remaining measurements were made in regions of low or intermediate concentrations of radon in water. Outdoor radon corresponded well with geology and in some instances was comparable with indoor radon concentrations. The average ambient radon concentration of 26 Bq m^{-3} reported by Hess and others (1982) is nearly twice the US average reported by Hopper and others (1991).

Site Studies

Long-term studies of outdoor radon have been conducted at a number of sites around the country, including Socorro, New Mexico (Wasiolek and others 1996; Wasiolek and Schery 1993); Fort Collins, Colorado (Borak and Baynes 1999); Chester, New Jersey (Fisenne 1988); suburban New Jersey and Central Park, New York (N. Harley, personal communication). Researchers from the New Mexico Institute of Mining made semicontinuous measurements of ^{222}Rn during the winter of 1991-1992 at a site 1 m above a golf course in Socorro, using a two-filter continuous monitor (Wasiolek and Schery 1993). Grab samples were also measured with a two-filter manual system, and meteorologic measurements were made 5 m above the ground. Both attached and unattached radon progeny were measured and an effective dose rate between 0.2 and 0.7 mSv y^{-1} was calculated. Radon concentrations varied over the period but generally were 5-10 Bq m^{-3} and had a geometric mean of 10.2 Bq m^{-3}. In a follow-on study, ^{220}Rn was measured at the same site from February 1994 to February 1995 (Wasiolek and others 1996); an average effective dose of ^{220}Rn decay products of 0.025 mSv was calculated.

From 1990 to 1997, continuous and passive monitoring of ambient radon at a site in northern New Jersey (N. Harley, personal communication) yielded an average of 8 Bq m^{-3}. Passive alpha-track detectors were measured seasonally and compared favorably with continuous (hourly) monitoring with a flowthrough scintillation counter. The Chittaporn and others (1981) continuous monitor used in the study is one of the few monitors that detects only radon and removes the decay products at formation with an electret. Decay products can introduce error into radon measurements that use flow-through scintillation counters. Monthly average outdoor radon concentrations were higher during the colder months of the year, however it is the daily variations of the outdoor radon that account for the variation by a factor of 2 in radon concentration. A site in Central Park, New York was also monitored for outdoor radon over a 3-y period using a passive alpha-track detector; they found an average of 7 Bq m^{-3} and trends similar to those in their northern New Jersey study (N. Harley, personal communication).

In a 3-y study (1993-1995) at a single site in Fort Collins, Colorado measurements (Borak and Baynes 1999) were made 1 m above the ground at 15-min intervals with a 1 liter flowthrough scintillation flask. Researchers found that outdoor radon responded to changes in temperature, wind speed, barometric pressure, and precipitation and varied seasonally and diurnally. Daily averages varied from 2.5 to 64.5 Bq m^{-3}, and monthly averages varied from 12 to 18 Bq m^{-3}; the magnitude of the variation in concentrations seems to be due to diurnal changes and specific meteorologic events and less to long-term seasonal response.

Fisenne (1988) measured outdoor radon in Chester, New Jersey with a continuous two-filter monitor for 9-y beginning in 1977. The annual average radon varied from 7.00 to 9.25 Bq m^{-3} and hourly measurements ranged from 0.4 to

TABLE 2.2 Statistical Summary of Outdoor-Radon Surveys in the United States (Bq m^{-3}).

Location	Time	Avg.	Geo. Mean	Geo. Std. Dev.	Range	Method	Reference
California (68 sites)	1 y	18.1	15.5	1.8	0.3-56.0	Etched track	Liu and others (1991)
Iowa (111 sites)	4+y	30.0	29.0	1.4	7.0-55.0	Etched track	Steck and others (1999)
Maine (51 sites)	6 mo.	27.0	17.6		2.9-160.0	Etched track	Hess and others (1982)
Minnesota (64 sites)	1 y	22.0	19.0	1.8	4.0-55.0	Etched track	Steck and others (1999)
Nevada (50 sites)	30 d	15.1	13.1	—	2.6-52.0	Electret ion chamber	Price and others (1994)
Socorro, NM	4 mo.	12.5	10.2	2.0	1.3-50.3	Continuous 2-filter	Wasiolek and Scherly (1993)
Fort Collins, CO	3 y	18.0	15.0	1.7	2.5-64.5 (daily)	Continuous monitor	Borak and Baynes (1999)
Suburban No. NJ	7 y	8.0	6.0	1.8	4.0-24.0 (daily)	Continuous monitor Etched track	N. Harley (personal communication)
Chester, NJ	9 y	8.1	—	—	0.4-63.0 (hourly)	Continuous 2-filter	Fisenne (1988)
Central Park, NY	3 y	7.0	—	—		Etched track	N. Harley (personal communication)

63 Bq m^{-3}. Again, diurnal changes in concentration and meteorologic events accounted for most of the variability in the hourly readings.

Statistical Analysis and Summary of Data

In all studies examined for this report, the spatial and temporal variation is striking and consistent and follows geologic, diurnal, meteorologic, and seasonal controls. However, comparison of the statistics from the studies (table 2.2) reveals that arithmetic and geometric means across time and geography were within the range of 6-30 Bq m^{-3}. Individual readings and hourly or daily averages were in the range of 1-63 Bq m^{-3}, with the exception of Maine. Both Iowa and Maine had higher average outdoor radon than the other areas, and the geometric mean for Iowa was significantly higher than all the others.

Average Ambient Radon

In the original charge to the committee, EPA requested "central estimates for a population-weighted average national ambient concentration for radon, with an uncertainty range. Comparisons of the contribution of radon in water to other sources of indoor radon will be made and comparisons will be made to outdoor levels." The charge was amended to include a discussion of alternatives to population-weighted averages and of spatial and temporal variation. The ambient-radon data of Hopper and others (1991) are the only data that provide some portion of national coverage over an extended period, but the committee has concerns about the appropriateness of using these data to develop a population-weighted average for the United States. Hopper and others contend that the data cannot be used to represent the ambient radon of the state that contains the sampling site. However, they do think that the ambient radon measured at each site is representative of that site. Using population data from the 1990 census, one can calculate a population weighted average (A_{PW}) by summing the products of each site's seasonal average (A_i) and the city population (P_i) and then dividing by the sum of the city populations.

That assumes that the ambient radon measured at the site is representative of the ambient radon of the city. The population-weighted average radon concentration is given by:

$$A_{PW} = \frac{\sum_{i=1}^{N=50} A_i \times P_i}{\sum_{i=1}^{N=50} P_i} \tag{2.1}$$

where N is the number of sites. The population-weighted average radon concentration is 14.0 Bq m^{-3}. The total population of the cities for all the sites in the national outdoor radon survey was about 24 million, or slightly less than 10% of the US population. The calculation is dominated by New York City, which has a large population and a lower than average ambient radon measurement, and therefore the population weighted average is less than the unweighted average of 14.8 Bq m^{-3}. Further, the sites were not chosen to be a statistical representation of the population across the United States, nor were the sites chosen to be representative of ambient radon within each state or sample the geology of the country. The overall distribution of the seasonal data for each site in the Hopper and others (1991) data is given in figure 2.8. The committee feels that it is more reasonable to recommend an (unweighted) arithmetic average radon concentration of

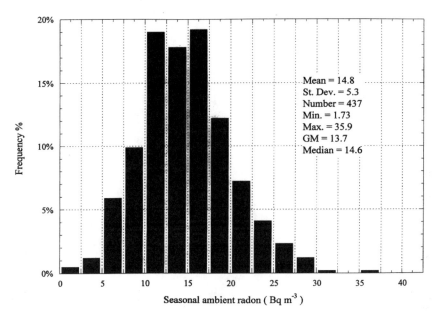

FIGURE 2.8   Frequency distribution of average seasonal ambient radon for the United States.

15 Bq m^{-3} for the United States and that it lies within a confidence interval of 14-16 Bq m^{-3}.

It is evident that radon concentrations in air, in water, and indoors vary systematically across the United States and that this variation should be part of any regional consideration of multimedia assessment and mitigation. A comprehensive, geographically based ambient-radon study that incorporates the major population areas of the United States and their geologic variability would provide the basis for a valid population-weighted ambient radon concentration. Focused regional studies of ambient radon in high-radon areas such as the glaciated northern tier of states and states of the Appalachian Mountains, Rocky Mountains, and Basin and Range would yield better information on overall exposure and more-realistic baseline information for evaluating the contribution of the ambient concentration to what is observed in indoor air.

3

Transfer of Radon from Water to Indoor Air

The estimation of the increment of airborne ^{222}Rn in a dwelling that arises from the use of water that contains dissolved radon is a complex problem. It involves the solubility of radon in water, the amount of water used in the dwelling, the volume of the dwelling, and the ventilation rate. The amount of waterborne radon escaping into the air is different throughout a dwelling but is higher in areas of active water use such as bathrooms and kitchens. It also depends on the radon concentration in the water and the activities that are taking place. However, it is common to estimate the average incremental concentration throughout the dwelling $(\overline{\Delta C_a})$ and derive a transfer coefficient as the ratio of that average concentration in water $\overline{C_w}$:

$$\text{Transfer Coefficient} = \frac{\overline{\Delta C_a}}{\overline{C_w}} \tag{3.1}$$

The relationship used to estimate the transfer coefficient has been derived by Nazaroff and others (1987) on the basis of the assumption that a house is a single well-mixed volume. The transfer function can then be described by:

$$\frac{\overline{\Delta C_a}}{\overline{C_w}} = \frac{We}{\lambda V} \tag{3.2}$$

where W is the time-averaged water use rate; e is the use-weighted transfer efficiency of ^{222}Rn from water to air; λ is the air-exchange rate, which is assumed to be uncorrelated with the water use rate; and V is the volume of the dwelling.

This chapter reviews published transfer-coefficient measurements. In addition, it performs a distributional analysis in the same manner as Nazaroff and others (1987) and compares the results.

MEASUREMENTS OF TRANSFER COEFFICIENTS

There are only a few measurements of transfer coefficients (from water to air) in the literature, and most of them refer to a limited number of geographic areas. Thus, there is considerable uncertainty in the extrapolation of the resulting distribution to the entire housing stock of the United States. The earliest reports of the measurements of the transfer coefficient were made by Gesell and Prichard (1980) and Castrèn and others (1980). They assumed that all the indoor radon was due to radon in the water, so the transfer coefficient was overestimated. Gesell and Prichard took measurements in apartments, where soil-derived radon is likely to contribute little to measured values. Castrèn's measurements, taken in houses in Finland, made the same assumption and are thus not likely to be useful. The actual values for the dwellings in Finland were not reported, and the data are no longer available (Castrèn, private communication, 1997).

McGregor and Gourgon (1980) measured C_a and C_w in six conventional residences, five trailers, and two schools in Nova Scotia, Canada. Only upper limits for the transfer coefficients can be estimated, because ventilation rates and water-use patterns were not determined. They found values of $0.032\text{-}0.24 \times 10^{-4}$ for the trailers and $0.038\text{-}0.52 \times 10^{-4}$ for the dwellings. The values are lower than many of the reported coefficients. Because the ventilation and water-use rates are not available, it is not possible to know whether the values are low as a result of high ventilation rates, low water use, or some combination of the two. Owing to the uncertainty in separating the contributions of soil gas and drinking water to the indoor radon concentrations, the transfer-coefficient values only for the five trailers have been included in the committee's estimated distribution of values.

During the late 1970s and throughout the 1980s, Hess and co-workers (1990; 1987b; 1987a; 1982) measured the transfer coefficients in a series of houses across Maine. In all, measurements were made in about 70 houses. They have used an approach that they term the "burst method." Measurements are made during a 2-h period when a series of water-use activities are performed. The idea is to use as much water during this 2-h period as would typically be used over a 24-h period. Radon is also monitored over a second 24-h period during which the residents use water according to their normal daily routine. The water-use activities during the second 24-h period were recorded. Ventilation rates were also measured directly.

As part of a study of the effectiveness of radon-mitigation methods (Deb 1992), concentrations of radon in air and water were measured before and after the initiation of a water treatment to reduce the concentration of radon in the water. Measurements of radon in air and water use were made in each of 119 houses before and after treatment. In two homes one measurement was unavailable; therefore, values for only 117 homes were used. In two of the communities, concentrations of radon in water were generally below 185,000 Bq m^{-3}; thus, the increment of airborne radon was small and difficult to measure accurately. In addition, not all of the before-and-after measurements were made in comparable seasons, so there might be substantial errors in many of the measurements. However, in a number of the homes, the measurement after treatment of the water was higher than the measurement before the reduction in waterborne radon. It is possible to estimate the transfer coefficient from the before-and-after radon concentration measurements appropriately weighted for the measured water use. However, some of the values are negative and therefore invalid. To incorporate all of the data into the distribution of values, the measurements for each home were averaged. Examination of the transfer coefficient as a function of the radon concentration in water before treatment suggested that choosing homes where the untreated-water concentrations were greater than 81,000 Bq m^{-3} would avoid most of the invalid results. Use of this criterion resulted in 31 values, which included only three negative ones.

Lawrence and others (1992) made a series of measurements in 29 homes in Conifer, Colorado. The air volume of each home and the volume of water used were determined, and the air and water ^{222}Rn concentrations were measured. However, ventilation rates were not measured, so the authors only estimated the minimum and maximum values of airborne radon resulting from release from the water used in the homes. Thus, minimum and maximum values of the transfer coefficients could be calculated from the results given in their paper. To use the resulting information in the overall distribution of measured transfer coefficients, a best estimate of the transfer coefficient was calculated by taking the square root of the product of the maximum and minimum values.

Chittaporn and Harley (1994) have measured the water-use contribution to airborne radon in an energy-efficient home in New Jersey. They estimate a transfer coefficient for the home at 1.7×10^{-5}.

The resulting distribution of values of the measured transfer coefficient is shown in figure 3.1. The median is 4.5×10^{-5}, and the average is 8.7×10^{-5} with a standard deviation of 1.2×10^{-4}. When plotted on a logarithmic scale, the central portion of the distribution is fairly linear and thus the geometric properties have also been calculated. The geometric mean is 3.8×10^{-5} with a geometric standard deviation (GSD) of 3.3.

MODELING OF TRANSFER COEFFICIENT

The model of Nazaroff and others (1987) is used here to estimate the central values of the distribution of transfer coefficient. As in that report, it is assumed that the underlying distributions of the house volumes, ventilation rates, water

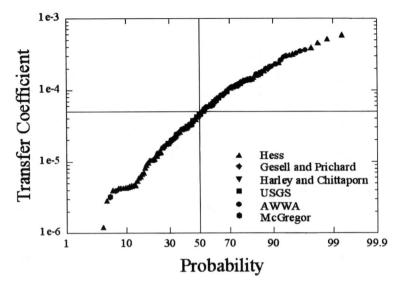

FIGURE 3.1   Cumulative probability distribution of measurements of transfer coefficient.

use, and use-weighted transfer efficiency are lognormal and so can be combined as done by Nazaroff and others (1987). There are new data on these input variables since the Nazaroff and others estimation, and they are summarized in the following sections.

House Volumes

A new survey of US dwelling volumes has recently been reported by Murray (1997). Two sets of values are reported; one is based on a 1993 survey of dwelling area conducted by the US Department of Energy (DOE) (1995), and the second is based on volumes obtained from a large number (over 4,000) of measurements of air-exchange rates using perfluorocarbon tracers between 1982 and 1987. Murray reports that the values fit a lognormal distribution well. Because the DOE survey represents a substantially larger number of homes (7,041), the geometric mean of 320 m^3 and a GSD of 1.8 derived from those data is used. The committee obtained the DOE database and extracted the area and the number of occupants per dwelling. Assuming a ceiling height of 2.44 m, house volumes per occupant were calculated. Figure 3.2 shows the distribution of dwelling volumes per occupant for each number of occupants and the overall distribution of dwelling volume per occupant. For the overall distribution, the geometric mean is 115 m^3 of volume per occupant with a GSD of 2.0.

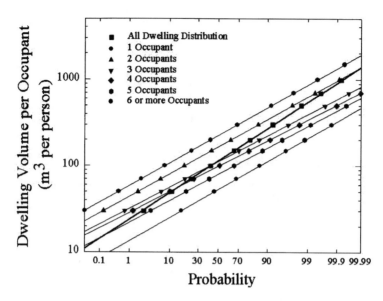

FIGURE 3.2   Cumulative probability distribution of volume of dwelling per occupant derived from 1993 DOE Residential Energy Consumption Study.

Ventilation Rates

Murray and Burmaster (1995) made a similar analysis of ventilation rates, based on the Brookhaven perfluorocarbon-tracer measurement database containing data on 2,844 dwellings. A geometric mean of 0.53 air changes per hour with a GSD of 2.3 was calculated. This value is somewhat lower than had been used by Nazaroff and others (1987). As Murray and Burmaster indicate in their paper, the houses used in the studies that constitute the database were not chosen on any statistical sampling basis. Most of the measurements were made in the northeastern United States or California and during the winter. Although the authors have made an effort to stratify the data and project the data to the entire United States, they may not have been able to fully represent the variability inherent in the US housing stock. The authors themselves indicate that they have concerns about the representativeness of their results. If the ventilation rate is negatively biased, its use in the model would tend to overestimate the transfer coefficient.

Alternatively, a modeling study by Sherman and Matson (1997) that used the larger DOE Residential Energy Consumption Study (DOE 1995) estimated a mean value of 1.1 air changes per hour. This study provided mean ventilation rates for each county in the country. Sherman and Matson estimate the number of houses of various types and sizes in each county in the United States by using data from the 1990 national census and the same DOE survey used to estimate house volumes. Sherman and Dickerhoff (1994) present the leakage rates of various types of US homes. Combining the data with average county weather, they calculate the distribution of residential ventilation rates for the population of US dwellings. The arithmetic and geometric means of these data are essentially the same: 1.1 air changes per hour with a GSD of 1.1. Thus, although the values might represent the central tendency in the data, they do not provide a good representation of the variability over housing types.

To combine the modeled ventilation rates into a transfer coefficient, it is necessary to be able to reflect the variability of the values for individual dwellings in the same way that the variability in water use, housing volume per occupant, and use-weighted transfer efficiency are reflected in their distributional parameters. Nazaroff and others (1987) looked at several smaller data sets of measured ventilation rates. They report geometric means of 0.90 and 0.53 air changes per hour and GSDs of 2.13 and 1.73, respectively. Those values are similar to the GSD of 2.3 estimated by Murray and Burmaster. Thus, it appears that a reasonable GSD for the underlying distribution of individual dwellings should be around 2. Because the Murray and Burmaster approach was to estimate the distribution over the entire housing population, the committee adopted their GSD to propagate into estimates of the geometric mean and standard deviation of the transfer coefficient. Using the Sherman and Matson model arithmetic mean and a GSD of 2.3 implies a geometric mean of about 0.77 air changes per hour. The committee also considered using the calculated geometric mean of 1.07 from the Sherman and Matson data with the higher GSD. That ventilation rate would yield a geometric mean transfer coefficient of 3.9×10^{-5} and an average 8.8×10^{-5}.

There is thus a disparity in the estimated geometric means between the available measurements and the results of modeling. The committee was concerned about the relatively few measurement data on ventilation in some areas of the country and in seasons other than winter. The model looks as though it might better represent the distribution of housing and the seasonality of ventilation rates across the United States. Clearly, neither source of information is fully satisfactory in providing the needed input to the estimation of the transfer coefficient.

Water Use Per Occupant

Residential water use in five cities was reported by Bowen and others (1993). The cities were Altamonte Springs, FL; Nashua, NH; Norman, OK; Portland, OR; and Tucson, AZ. Those authors reported average per capita use in two seasons and as a function of flow rates. There was no separation of indoor use from outdoor uses, such as watering lawns and filling pools. Only water used indoors will contribute to the indoor radon concentration, so these data are not helpful unless it is possible to estimate the indoor use fraction, and there is not direct information available to do so. Thus, the data are not directly useful in estimating water use per occupant without assuming the fraction of indoor water use.

However, additional data from an ongoing residential water-use measurement program were made available by the American Water Works Association (DeOreo, private communication, 1997). This study has provided information on detailed daily amounts of use for a number of specific activities (clothes-washing, dishwashing, toilets, showering, baths, faucets, etc.) for 595 houses and permits a determination of the indoor use rates. It was conducted in Boulder; CO, Denver, CO; Eugene, OR; Seattle, WA; and San Diego, CA. Systems were installed in the homes to provide a log of each water use, its duration, and the total volume of water used. For some of the days being monitored, water use was extremely low, suggesting that the occupants were not home and so were not using water in the normal manner. By carefully reviewing the various records, it is possible to eliminate those values from the database. That results in the distribution observed in figure 3.3. Although there was a wide variation in total water use among the locations, there was much less variation among the indoor use rates; the average was 0.28 ± 0.20 m^3 per person per day, and the geometric mean was 0.23 with a GSD of 1.8.

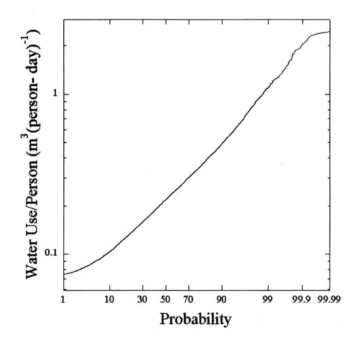

FIGURE 3.3   Cumulative probability distribution of water use per day per person based on data from Boulder, CO; Denver, CO; Eugene, OR; Seattle, WA; and San Diego, CA.

Use-Weighted Transfer Efficiency

The transfer efficiency is the fraction of the radon in the water that is released to the air during the activities that use water (showering, clothes-washing or dishwashing, and so on). The earliest values of the transfer efficiencies have been previously reported by EPA (Becker III and Lachajczyk 1984) and Nazaroff and others (1987). The transfer efficiency of shower heads has been measured by Fitzgerald and others (1997) and by Bernhardt and Hess (1996), who found values quite similar to those used by Nazaroff and others. The transfer efficiency for each activity was combined with the water use data to yield a use-weighted transfer coefficient. A geometric mean of 0.52 with a GSD of 1.3 was obtained.

Transfer-Coefficient Model Results

The new input values and the earlier values of Nazaroff and others are compared in table 3.1. Both the dwelling volume and the water use per occupant are greater in the committee's analysis. The transfer efficiency has remained essentially the same. The resulting geometric mean value of the transfer coefficient has risen to 0.55×10^{-4} with a GSD of 3.5. The arithmetic average transfer coefficient is estimated at 1.2×10^{-4}; it is considerably higher than the geometric mean value because of the high degree of skewness of the distribution.

CONCLUSIONS

There is reasonable agreement between the average value of the transfer coefficient estimated by the model and the value calculated from the measured data.

TABLE 3.1   Lognormal Distributions of Parameters in Transfer-Coefficient Calculation

Parameter	Nazaroff and others (Nazaroff and others 1987)		Committee	
	Geometric Mean	Geometric Standard Deviation	Geometric Mean	Geometric Standard Deviation
Dwelling volume per occupant (m^3 person^{-1})	99	1.9	115	2.0
Ventilation rate	0.68	2.0	0.77 or 1.07	2.3
Transfer efficiency water to air	0.55	1.1	0.52	1.3
Water use per capita (m^3 person^{-1} h^{-1})	7.9×10^{-3}	1.6	9.4×10^{-3}	1.8
Geometric mean transfer coefficient	6.5×10^{-5}	2.8	5.5×10^{-5} or 3.9×10^{-5}	3.5

The average of the measurements was 8.7×10^{-4} with a standard error of 1.0×10^{-4}. With a modeled geometric mean ventilation of 1.07 air changes per hour, the transfer coefficient is calculated to have the same value as the measurements. However, if the committee uses the estimate of the geometric mean of the ventilation rate of 0.77, the resulting estimate of the transfer coefficient is 1.2×10^{-4}. The committee feels that there are problems with both the measurements and the model results. Considering the problems with both the measurements of the transfer coefficient and the measurements that are the input values for the model, the committee recommends that the Environmental Protection Agency continue to use 1.0×10^{-4} as the best central estimate of the transfer coefficient, based on the available data.

Because of the uncertainty in the value of the ventilation rate and its distributional characteristics, the committee recommends assuming that the transfer coefficient is between 0.8 and 1.2×10^{-4}. The committee is not assigning a specific uncertainty to the central estimate, but rather assumes that the central estimate has the highest likelihood of lying within that range.

4

Dosimetry of Ingested Radon and its Associated Risk

Estimates of dose to various tissues of the body from ingestion of radon dissolved in drinking water and the resulting health risks are developed in this chapter. A review of the literature indicated a wide range in the reported dose per unit intake (dose coefficient), and neither national nor international radiation-protection agencies have provided authoritative values. In particular, values reported for the dose to the stomach per unit radon activity ingested (a dose coefficient) vary widely, are often based on assumptions that are not documented, and often are not based on contemporary dosimetric methods. The central issues are: 1) the extent to which radon diffuses into the wall of the stomach, and 2) the behavior of radon and its decay products in the body. Studies of the behavior in the body of inhaled and ingested radon indicate that radon is readily absorbed by blood and is rapidly eliminated from the body in exhaled air. Because of the wide range in dose coefficients reported in the literature, the committee has undertaken an independent dosimetric analysis using the methods of contemporary radiation dosimetry.

The chapter begins with a brief review of the relevant physiochemical properties of radon, the consumption of drinking water, and estimates of dose and risk reported in the literature. Following these introductory discussions the committee's estimates of dose and risk are presented.

INTAKES AND CONSUMPTION OF WATER

In regulating other drinking water contaminants (EPA 1994b) the US Environmental Protection Agency currently uses the quantity 2 L per day for adults and 1 L per day for "infants" (individuals of 10 kg body mass or less) as default drinking water intakes (EPA 1994b). The combined mean value of 1.2 L d^{-1} consists of both direct use, such as tapwater ingestion as well as indirect use, such as juices and other beverages that contain tapwater, such as coffee. A National Research Council committee (1977) has suggested that daily consumption of water can vary with extent of physical activity and fluctuations in temperature and humidity and that people who live in warmer climates might have higher intakes of water.

Numerous studies have developed data on drinking-water intake. All the studies that are available were based on short-term survey data. One of the more commonly cited studies on water intake is the Ershow and Cantor (1989) study. They estimated water intake on the basis of data collected by the US Department of Agriculture 1977-1978 Nationwide Food Consumption Survey and calculated daily intake and total water intake by various age groups of males, females, and both sexes combined. They defined tapwater as "all water from the household tap consumed directly as a beverage or used to prepare foods and beverages" and defined total water intake as tapwater plus "water intrinsic to foods and beverages." Table 4.1 summarizes data from the Ershow and Cantor study.

The combined mean value of 1.2 L d^{-1} (table 4.1) is for all uses of tapwater, which consists of both direct use (i.e., direct ingestion) and indirect use, i.e. making coffee, tea, etc. As noted by the EPA (1994b), the concern about radon dissolved in water is largely for the water that is ingested directly. The EPA has estimated that slightly more than half of the tapwater use is directly ingested. The committee has adopted a value of 0.6 L d^{-1} for direct use. This value is similar to that used by the EPA in their Multimedia Risk Assessment (EPA 1994b). However, the committee has conservatively assumed for direct use, that all of the radon in the tapwater remains dissolved in the process of transferring the water from the tap to the stomach.

PHYSICOCHEMICAL PROPERTIES OF RADON

Radon, a noble gas, is essentially chemically inert. Unlike the other noble gases, radon has no known stable isotope. Rather it has 36 radioactive isotopes and isomers, which range in mass number from 198 to 228. The radon isotope of interest here is ^{222}Rn (physical half-life, 3.825 d), a member of the decay series beginning with the primordial radionuclide ^{238}U. ^{222}Rn emits alpha particles as it spontaneously decays to a series of short-lived radioactive decay products, which are followed by a longer-lived series headed by ^{210}Pb (half-life, 22.3 y), as shown earlier in figure 1.1. The cumulative energies of the radiation emitted by the members of the decay series (alpha particles, electrons, and photons) are shown in table 4.2. The tabulated values represent the average or expected energy of the indicated radiation emitted per atom of ^{222}Rn initially present. The entry for a particular member includes the contribution of the member and its precursors

TABLE 4.1   Tapwater Intake by Sex and Age[a]

Age Group (y)	Daily Tapwater Intake (mL)					
	Mean	SD	Median	90% UCL[b]	95% UCL[b]	99% UCL[b]
Males						
<0.5	250	232	240	569	757	NA
0.5 to 0.9	322	249	264	634	871	NA
1 to 3	683	406	606	1228	1464	2061
4 to 6	773	414	693	1336	1530	1900
7 to 10	802	437	738	1391	1609	2055
11 to 14	970	547	877	1714	2019	2653
15 to 19	1120	644	1019	1974	2283	3090
20 to 44	1354	788	1216	2309	2837	4065
45 to 64	1633	783	1510	2650	3094	4213
65 to 74	1594	719	1457	2502	2812	NA
>75	1517	667	1443	2332	2696	NA
All ages	1250	759	1123	2205	2673	3760
Females						
<0.5	293	259	240	672	800	NA
0.5 to 0.9	333	281	278	712	759	NA
1 to 3	606	368	532	1114	1339	1806
4 to 6	709	395	622	1231	1491	1932
7 to 10	772	395	726	1299	1475	1888
11 to 14	881	490	797	1531	1814	2382
15 to 19	883	513	800	1565	1839	2452
20 to 44	1182	634	1089	1996	2323	3132
45 to 64	1483	670	1394	2303	2668	3666
65 to 74	1429	603	1360	2247	2561	3082
>75	1300	540	1250	1998	2242	2933
All ages	1147	648	1049	1988	2316	3097
Males and Females Combined						
<0.5	272	247	240	640	800	NA
0.5 to 0.9	328	265	268	688	764	NA
1 to 3	646	390	567	1162	1419	1899
4 to 6	742	406	660	1302	1520	1932
7 to 10	787	417	731	1338	1556	1998
11 to 14	925	521	838	1621	1924	2503
15 to 19	999	593	897	1763	2134	2871
20 to 44	1255	709	1144	2121	2559	3634
45 to 64	1546	723	1439	2451	2870	3994
65 to 74	1500	660	1394	2333	2693	3479
>75	1381	600	1302	2170	2476	3087
All ages	1193	702	1081	2092	2477	3415

[a]Data from Ershow and Cantor (1989).
[b]UCL = upper confidence limit.

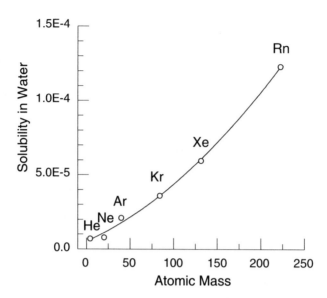

FIGURE 4.1   Solubility of the noble gas elements in water (CRC 1996) at body temperature, shown as a function of atomic mass. The solubility is expressed as the mole fraction
of the gas in the mixture.

(listed above it), including ^{222}Rn. The kinetic energy of the emitted alpha particles for the ^{222}Rn series (24.5 MeV) accounts for 89% of the total emitted
energy (27.6 MeV). A substantial fraction (78%) of the alpha energy is associated
with the short-lived radon decay products (19.2 of 24.5 MeV). If an atom of ^{222}Rn
entered the body, in the absence of any biologic removal mechanisms for it or its
decay products, the energies listed in table 4.2 would be available for deposition
within the tissues of the body. However, ingested and inhaled radon is known to
be promptly removed from the body by exhalation. Biologic removal processes
are also applicable to the decay products formed within the body, but the short
half-life of some decay products limits the importance of these removal processes. The decay products formed within the body may enter their own metabolic pathways and routes of excretion from the body. Ingested radon is removed
from the body through exhalation while the longer-lived decay products are
eliminated by urinary and fecal excretion.

   The extent to which radon is absorbed from the gastrointestinal (GI) tract and
retained in the body is determined, in part, by its solubility in blood and in the
tissues. The solubility of the various noble gases in water (CRC 1996) at body
temperature is shown graphically as a function of atomic mass in figure 4.1.
Radon is considerably more soluble in water than the lighter noble gases—about
15 times as soluble as helium and neon. Data on solubilities of the noble gases in
body tissues exhibit a similar relationship although the data are more variable.

TABLE 4.2   Cumulative Energy of Radiations Emitted in the Decay of ^{222}Rn and Members of Its Decay Series

Nuclide	$T_{1/2}$	Alpha	Electron	Photon	Total
			Energy (MeV per ^{222}Rn atom)		
Rn-222	3.8235 d	5.49	—	0.000399	5.49
Po-218	3.05 m	11.5	—	0.000408	11.5
Pb-214	26.8 m	11.5	0.293	0.250	12.0
Bi-214	19.7 m	11.5	0.952	1.76	14.2
Po-214	164.3 μs	19.2	0.952	1.76	21.9
Pb-210	22.3 y	19.2	0.990	1.76	22.0
Bi-210	5.01 d	19.2	1.38	1.76	22.3
Po-210	138.38 d	24.5	1.38	1.76	27.6

Radon is readily absorbed from the GI tract and distributed among the tissues, in part because of its relative solubility in blood and in tissue. The ratio of solubility in tissue to that in blood is referred to as the partition coefficient. Measurements of the solubilities and partition coefficients of argon, krypton, xenon, and radon have been reported. Considerable data are available on xenon because of its use in assessing blood flow. Data on radon are less plentiful; the work of Nussbaum (1957) is their major source. Data on the partition coefficients of krypton, xenon, and radon are summarized in table 4.3. Of particular note are the higher partition of radon in blood (7 times that of krypton) and its higher partition in adipose tissue. Adipose tissue is the major tissue of deposition of radon that has entered the systemic circulation.

Estimates of Dose from Ingested Radon

The inhalation hazard of radon and its short-lived decay products has long been of concern in occupational radiation protection and public health. Ingestion

TABLE 4.3   Partition Coefficients of Noble Gases

Organ	Krypton	Xenon	Radon
Blood/air	0.06	0.18	0.43
Adipose tissue/blood	5.50	8.00	11.2
Muscle/blood	1.09	0.70	0.36
Brain/blood	1.13	0.75	0.72
Kidney/blood	—	0.65	0.66
Testes/blood	—	—	0.43
Liver/blood	—	0.70	0.71
Bone/blood	—	0.41	0.36
Lung/blood	—	0.70	0.70[a]
GI-tract/blood	—	0.81	0.70[a]
Other/blood	—	—	0.70

[a]Default values used in analysis.

of radon has received considerably less attention from radiation protection agencies who are primarily concerned with occupational exposures, in part because ingestion intakes are readily avoided in the workplace. Thus, recommendations of the International Commission on Radiological Protection (ICRP) and the National Council on Radiation Protection and Measurements (NCRP) have not included guidance for the control of ingested radon. In the absence of such guidance a number of investigators have undertaken dosimetric and risk assessments of ingested radon.

The fate of radon in the body has been the subject of several investigations. In 1951, Harley and others (1994; 1958) examined the elimination of radon from the body in a series of measurements of radon in exhaled air after chronic inhalation. Hursh and others (1965) investigated the fate of ingested radon and proposed a concentration limit on radon dissolved in water. Von Doebeln and Lindell (1964) investigated the retention of ingested radon in the body. Brown and Hess (1992) investigated the transfer and kinetics of ingested radon. The retention in the body of the noble gases argon, krypton, and xenon were studied by Tobias and others (1949), Ellis and others (1977), Susskind and others (1977; 1976), and Bell and Leach (1982). The results of those investigations support a number of general observations:

- Ingested radon is absorbed from the gut.
- Exhalation is the major route of elimination from the body.
- Ingested radon is largely eliminated within an hour.
- Body adipose tissue is the major site of long-term retention.

The absorption and retention of inert gases in human body tissues have been extensively studied by several authors, including: Smith and Morales (1944), Morales and Smith (1944), Kety (1951), Bernard and Snyder (1975), Bell and Leach (1982), Palazzi and others (1983), Peterman and Perkins (1988), Harley and Robbins (1994), Sharma and others (1996). *In vitro* studies have provided data on the solubilities and partition coefficients of the noble gases and other chemically inert substances in human blood, adipose tissue, and individual tissues; the data have been summarized by Steward and others (1973).

Hursh and others (1965) derived a value for the maximum permissible concentration of ^{222}Rn in water on the basis of limiting the dose to the stomach. They assumed that radon diffuses through the stomach wall and enters the splanchnic blood flowing to the liver. The concentration of radon in the stomach wall was taken, conservatively, to be equivalent to that in the stomach contents. Von Doebeln and Lindell (1964) used the data of Hursh and others to estimate the dose to the stomach. More recently, Crawford-Brown (1991; 1989) estimated the dose to the stomach using a linear radon concentration profile in the wall. The dose to other organs was based on kinetics inferred from measurements of retention of ingested ^{133}Xe (Correia and others 1987). Harley and Robbins (1994) used a compartment model (the parameter values are not given in the paper) for the distribution of radon in the body based on data from Hursh and others (1965), Harley and others (1994), and Harley and Robbins (1994). They assumed that a fraction of the ingested radon diffuses into the stomach wall but that the vascular structure in the mucosa intercepts the radon before it reaches a depth from which the alpha emissions could irradiate the stem cells (N. Harley personal communication, see appendix A). Sharma and others (1996) used the compartment model of Peterman and Perkins (1988) and the data of Brown and Hess (1992) to estimate the dose from dissolved radon. They make no statements regarding the diffusion of radon into the stomach wall and computed, in a most unusual manner, the stomach dose as the alpha energy emitted within the stomach contents divided by the mass of the stomach wall and contents (C.T. Hess personal communication). The estimates of the dose to the stomach obtained by the various investigators are listed in table 4.4. All authors assumed that the short-lived decay products of radon decayed at the site of the ^{222}Rn decay; that is, the alpha energy of 19.2 MeV (see table 4.1) was associated with each ^{222}Rn decay. Those specifically considering the diffusion of radon within the stomach wall generally associated the first two alpha emissions, 11.5 MeV, with the radon decay in the stomach wall.

The data in table 4.4 indicate that the estimated dose to the stomach depends on the extent to which the investigators considered diffusion as a mechanism by which radon comes into intimate contact with the stomach wall; the highest dose coefficient is about 200 times the lowest (Sharma and others 1996; Harley and Robbins 1994; Brown and Hess 1992; Crawford-Brown 1989; Suomela and Kahlos 1972; Hursh and others 1965; Von Doebeln and Lindell 1964). Except for Harley and Robbins (1994), none of the investigators identified any basis for their assumption regarding the movement or lack of movement of radon into the stomach wall. Harley and Robbins (1994) assumed that the absorption of radon follows that of water which is predominantly from the small intestine, and cited the large countercurrent flow of fluid (1500 mL per day) from the stomach wall as

TABLE 4.4 Summary of Estimates of Equivalent Dose to Stomach per Unit Activity of ^{222}Rn Ingested

Authors	Diffusion	Dose Coefficient, Sv Bq^{-1}
Hursh and others (1965)	Yes	1.1×10^{-7}
Von Doebeln and Lindell (1964)	Yes	1.1×10^{-7}
Suomela and Kahlos (1972)	Yes	1.3×10^{-7}
Crawford-Brown (1989)	Yes	3.0×10^{-7}
Brown and Hess (1992)	See footnote[a]	8.8×10^{-8}
Sharma and others (1996)	See footnote[a]	8.2×10^{-8}
Harley and Robbins (1994)	Yes	1.6×10^{-9}

[a]Dose averaged over the mass of the stomach wall and contents.

evidence for lack of significant wall transport. No experimental evidence was provided by any of the investigators as a basis for their assumptions concerning the movement or lack of movement of radon into the stomach wall.

Behavior of Radon in the Body

The investigations by Harley and colleagues (1994), Harly and Robbins (1994) and Hursh and others (1965) provide basic information on the behavior of radon in the body. Harley and coworkers fit their observations to a function involving five exponential terms and associated the terms with tissue compartments. Bernard and Snyder (1975) interpreted the Harley data using a mammillary model to derive their estimates of the distribution of radon among the tissues. It is possible to observe the tissue distribution of noble gases in medical studies using radioisotopes of krypton and xenon. Correia and others (1987) used nuclear-medicine instrumentation to observe the behavior of ingested ^{133}Xe and then attempted to infer the fate of ingested radon. Presently, it is considered that compartment models describing the movement of a contaminant within the body which are consistent with anatomic and physiologic principles provide the best basis for interpretation of experimental observations. Models based on physiologic principles are referred to as physiologically based pharmacokinetic (PBPK) models. Details regarding this modeling practice are given in a review by Bischoff (1986).

The committee has used a PBPK model of ingested radon that was formulated using the blood-flow model of Leggett and Williams (1995) (see also Leggett and Williams 1991; Williams and Leggett 1989). The resulting model, shown in figure 4.2, is discussed in detail in appendix A. Briefly, radon is distributed by the blood flow to the organs, where its transfer depends on its solubility in tissue relative to that in blood—the partition coefficient. The blood volume of the body is apportioned among a number of compartments, which represent various blood pools. In figure 4.2, the compartment "Large Veins" represents the venous blood return from the systemic tissues, "Right Heart" and "Left Heart" the content of the heart chambers, "Pulmonary" the blood-exchanging gases in the lung, and "Large Arteries" represents the arterial blood flow to the systemic tissues. The compartment labeled "Gut Cont" in figure 4.2 is expanded in figure 4.3, where the GI tract is divided into four segments, the compartment "St Contents," representing the contents of the stomach, "SI Contents" that of the small intestine, "ULI Contents" that of the upper large intestine, and "LLI Contents" that of the lower large intestine. "St Wall," "SI Wall," "ULI Wall," and "LLI Wall" represent the walls of those segments. Ingested radon enters the stomach and is absorbed from the gut as indicated in the upper right of figure 4.3. In figures 4.2 and 4.3, dashed arrows denote the transfer of radon as a gas, and solid arrows correspond to the flow of radon dissolved in arterial (thicker arrows) and venous blood. As shown in figure 4.3, the gut is perfused by arterial blood,

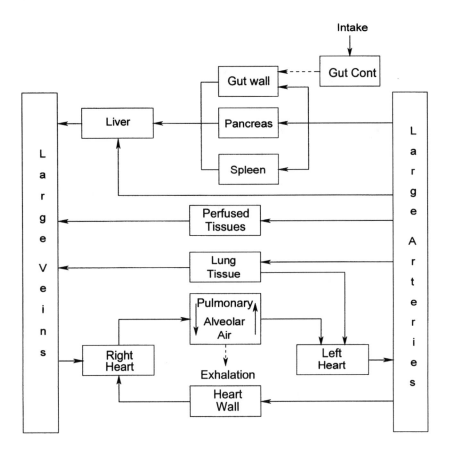

FIGURE 4.2   Diagram of PBPK model used for ingested radon (see Appendix A for details).

which, with blood flowing from the spleen and pancreas, enters the portal circulation to the liver as shown in figure 4.2. Venous blood is pumped by the right side of the heart to the pulmonary region of the lung, the "Pulmonary" compartment of figure 4.2, where radon dissolved in blood exchanges with alveolar air and is exhaled.

Ingested ^{222}Rn is readily absorbed and appears promptly in exhaled air. The studies of ingested radon have indicated that retention in the body is somewhat greater when radon is ingested with food (Brown and Hess 1992; Hursh and others 1965). The increased retention is presumably a result of the slower transfer of foodstuffs from the stomach to the small intestine. Whereas the small intestine is the major site of absorption of most nutrients, fractional amounts of some

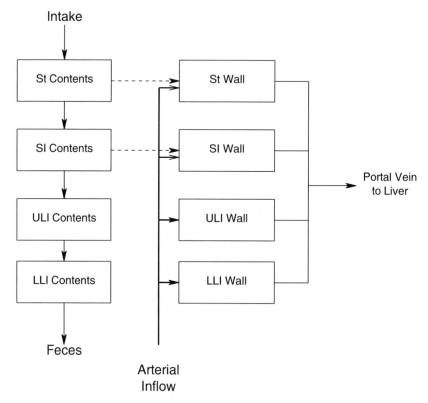

FIGURE 4.3   Expansion of the gut compartments of figure 4.2 to show the four segments of the gastrointestinal tract.

materials, such as water, aspirin, and alcohol, are known to be absorbed from the
stomach.

   Further details regarding the radon biokinetic model, including the numerical values of the transfer coefficients of the resulting differential equations, are
given in appendix A. The equations are solved by assuming that a unit activity
(1 Bq) of ^{222}Rn is present in the stomach contents at time zero. The fractions of
ingested radon that remain in the body (in the contents of GI tract and in systemic
tissues) at various times after intake are shown in figure 4.4. The fraction of the
initial activity residing in various tissues as a function of time is shown in figure
4.5. The high radon uptake in the liver shown in figure 4.5 is a direct reflection of
the fact that all radon absorbed from the GI tract flows in blood from the GI tract
walls to the liver. The importance of adipose tissue as a site of deposition and
retention can be seen at later times; beyond about 30 min it is the major site of
radon deposition in the body.

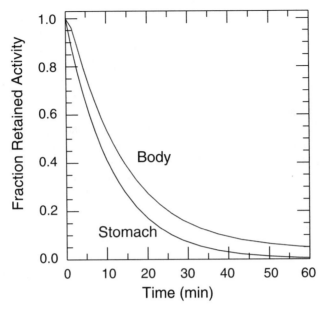

FIGURE 4.4   The fraction of ingested radon remaining in the body (with the contents of the gut and in systemic tissues) at various times following an intake by ingestion.

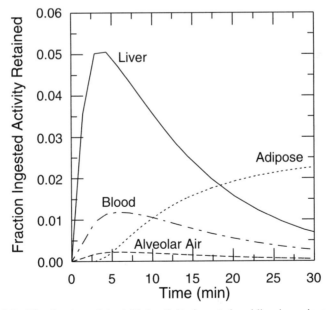

FIGURE 4.5   The fraction of the initial activity ingested residing in various tissues of the body as a function of time following the intake.

The PBPK model's predictions of radon retention in the body are compared with the observations of Hursh and others in figure 4.6. The solid line in figure 4.6 is the PBPK prediction, and the dashed lines are the fits of Hursh and others to their data on two male subjects who ingested radon on two occasions. The upper dashed line represents the retention observed shortly after a subject had ingested a breakfast that included "heavy whipping cream." The other observations were at least 2 h after a "normal light" breakfast. The comparisons in figure 4.6 were limited to the time period of the observations. The higher retention after the high-fat meal is consistent with the general observation noted above and presumably reflects a slower emptying of the stomach.

Figure 4.7 compares the PBPK model's prediction of radon retention with that of the mammillary model of Bernard and Snyder (1975) based on the Harley and others (1994; 1958) data on washout of inhaled radon. Also shown in the figure is the retention indicated by Crawford-Brown (1989) based on his analysis of the Correia and others (1987) data. The differences between the model predictions appear to be much as expected; for example, the Bernard and Snyder model might be expected to underestimate the retention because it was derived

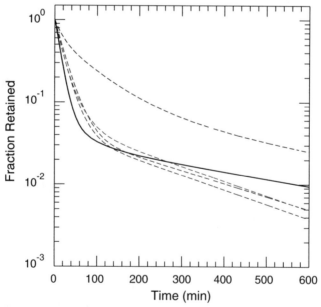

FIGURE 4.6   The PBPK model's predictions of radon retention in the body are compared to the observations of Hursh and others (1965). The solid line of figure 4.6 is the PBPK predictions and the dashed lines are the fits of Hursh et al. to their data for two male subjects who ingested radon on two different occasions.

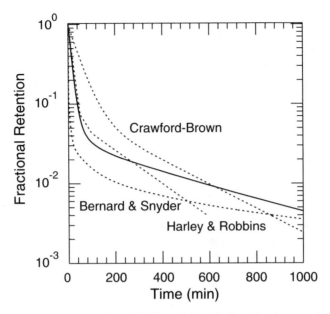

FIGURE 4.7 A comparison of the PBPK model predictions for the retention of radon with that indicated by the mammillary model of Bernard and Snyder (1975) based on the Harley and others (1994) data on the washout of inhaled radon. Also shown in the figure is the retention indicated by Crawford-Brown (1989) based on his analysis of the Correia and others data (1987) and the results of Harley and Robbins (1994).

from inhaled radon. The PBPK model appears to provide reasonable predictions of retained radon, and its physiological basis provides the basis of the distribution of radon among the organs of the body.

Diffusion of Radon in the Stomach

As seen from table 4.4, the dose to the stomach depends strongly on whether radon is considered to move into the stomach wall, presumably by diffusion. Alpha particles emitted within the contents of the stomach cannot penetrate the mucus layer lining the epithelium and cannot reach the stem cells at risk (the range of alpha particles in tissue is about 50-60 µm). This mucus layer is thought to be important in minimizing the exposure of the stomach epithelium to the acidic environment of the gastric lumen (Livingston and Engel 1995) and possibly acts as a barrier to the gastric absorption of drugs (Larhed and others 1997). The layer is composed primarily of mucin molecules that are continuously secreted and degraded by the gastric acids; although it is about 95% water, it substantially slows the diffusion of acid relative to water alone. It has been hypothesized that the continual movement of water and new mucin molecules away from the epithelial surface also carries acid away from the epithelium by convection (Livingston and Engel 1995). The interplay of these factors has been mathematically modeled (Engel and others 1984), but many of the physiological parameters of the model have not been measured.

In view of the importance of the diffusion mechanism in estimating the dose to the stomach, the committee found it useful to formulate a model within which it could investigate this mechanism. The model consisted of a spherical representation of a stomach with a volume of 250 mL. A mucus layer 50 μm thick was assumed. This was followed by a layer of surface cells 50 μm thick. The stem cells were considered to be distributed throughout a layer of tissue 200 μm thick. Below this layer, diffusion into capillaries was assumed to remove radon and reduce the concentration to zero. The concentration of radon in the contents of the stomach, assumed to be well mixed, was taken to decline exponentially with a half-time of 20 min. The model and its results are discussed in detail in appendix B. When the above parameters were used with a radon diffusion coefficient for the gastric wall of 5×10^{-6} cm^2 s^{-1}, the time-integrated concentration of radon at the depth of the cells at risk (200 μm) was found to be 30% of the time-integrated radon concentration in the contents of the stomach. The time-integrated concentration was found to be insensitive to the value assumed for the diffusion coefficient and to depend somewhat on the depth to which radon was assumed to diffuse.

Although the diffusion model of the stomach does not permit definitive conclusions, it does suggest that both radon concentration and its time integral vary over a rather limited range for a wide range in the diffusion coefficient. If the mucus layer is a barrier to radon diffusion, concentration in the wall could be substantially reduced. The chemical composition of the layer (95% water, degraded mucin, and soluble polymeric mucin secreted by the mucosa) does not suggest a strong diffusion barrier to inert substances like radon. The concentration of radon reached in the wall is controlled by the blood flow through the gastric mucosa, and the depth of microvasculature may be of considerable importance. The influence of the microvasculature of the small intestine on absorption of gases has been investigated (Bond and others 1977), but little information is available on the stomach. Further studies clearly are needed to determine the influence of the mucus layer and the capillary structures on the concentration of radon in the stomach wall. It should be noted that because the PBPK model cannot fully adhere to the microvasculature which removes most radon directly to the blood before it can diffuse near stem cells, the model is a conservative model.

The calculations of dose and risk reported below assume that the time-integrated concentration of radon at the depth of the stem cells is 30% of that in the lumen. The latter value, derived with the diffusion model of appendix B, is the basis of the committee's recommendation regarding the risks posed by radon dissolved in water. For comparative purposes, dose coefficients were calculated as bounding cases corresponding to the situations of table 4.4; that is, the assumptions that radon does not diffuse into the stomach wall and that concentration in the wall is the same as that in the stomach contents represent the two limiting cases.

FATE OF RADON DECAY PRODUCTS IN THE BODY

The members of the decay series through ^{214}Po are referred to as the short-lived decay products relative to the long-lived series headed by ^{210}Pb. The half-life of the "short-lived" decay product ^{214}Pb (26.8 min) is not short relative to physiological processes, inasmuch as, for example, during this half-life, blood passes through the heart more than 30 times. Thus, it is reasonable to assume that the ^{214}Pb has its own fate within the body, a fate that is distinct from that of radon. The calculations performed here include explicit consideration of the fate of each decay product in the manner of recent ICRP publications (1989; 1988); for further details, see appendix A.

Dose Coefficients for Ingestion of Dissolved Radon

The dosimetric analysis presented here is based on the current ICRP method (ICRP 1989), which is consistent with the schema of the Medical Internal Radiation Dose Committee (MIRD) of the US Society of Nuclear Medicine (Loevinger and others 1988). Both ICRP and MIRD consider the mean absorbed dose to a target region as the fundamental dosimetric quantity. The mean absorbed dose in the target region is relevant to cancer induction to the extent that it is representative of the dose to the cells at risk. If the cells at risk are not uniformly distributed within the target region or if the stochastic nature of the energy deposition is such that mean values are of questionable validity, it might be necessary to address the stochastic nature of irradiation.

The ICRP method considers two sets of anatomic regions. The set of "source regions" specifies the location of radionuclides in the body, and the set of "target regions" consists of organs and tissues for which the radiation doses are to be calculated. The source regions are those anatomical regions involved in the behavior of the radionuclide (and subsequent decay products) within the body. It is assumed that the radionuclide is uniformly distributed within the volume of the source region.

The mean energy absorbed in the target region depends on the types of the radiations (including their energies and intensities) emitted in the source regions, the spatial relationships between the source and target regions, and the nature of tissues between the regions. The details of these considerations are embodied in a radionuclide-specific coefficient called specific energy, or SE. For any radionuclide, source region S, and target region T, the specific energy at age t is defined as

$$SE(T \leftarrow S;t) = \frac{1}{M_T(t)} \sum_i Y_i E_i AF_i(T \leftarrow S;t) \tag{4.1}$$

where Y_i is the yield of radiation of type i per nuclear transformation, E_i is the average or unique energy of radiation type i, AF_i $(T{\leftarrow}S;t)$ is the fraction of energy emitted in source region S that is absorbed within target region T at age t, and $M_T(t)$ is the mass of target region T at age t. Age dependence in SE arises from the changes with age in both the absorbed fraction and the mass of the target region. The quantity AF_i $(T{\leftarrow}S;t)$ is called the absorbed fraction (AF) and, when divided by the mass of the target region, M_T, is called the specific absorbed fraction (SAF). Information on the energies and intensities of the radiation emitted by the members of the radon series is tabulated in ICRP Publication 38 (1983).

The SE values used here were computed with the SEECAL code of Cristy and Eckerman (1993). The calculations use files (electronic libraries) of the nuclear-decay data, SAFs for the emitted radiation, and values for the masses of the organs in people of various ages. The nuclear-decay data files and SAFs are those now used by ICRP (Cristy and Eckerman 1993; 1987). Organ masses for adults are taken from ICRP Publication 23 (1975). For children, age-specific organ masses are taken from Cristy and Eckerman (1987).

The absorbed-dose rate in target region T includes contributions from each radionuclide in the body and from each region in which radionuclides are present. The absorbed-dose rate, $\dot{D}_T(t, t_0)$, at age t in region T of a person of age t_0 at the time of intake, can be expressed as:

$$\dot{D}_T(t,t_0) = c \sum_s \sum_j q_{s,j}(t) SE(T \leftarrow S;t)_j \tag{4.2}$$

where $q_{s,j}(t)$ is the activity of radionuclide j present in source region S at age t, $SE(T{\leftarrow}S;t)_j$ is the specific energy deposited in target region T per nuclear transformation of radionuclide j in source region S at age t, and c is any numerical constant required by the units of q and SE. The absorbed dose is the time integral of the absorbed-dose rate.

The equivalent dose is the absorbed dose of the various kinds of radiation weighted by a factor that represents their relative contributions to the biologic insult. The weighting factor, referred to as the radiation weighting factor (earlier called the quality factor), represents a judgment of the relative biologic effectiveness of the different radiations (ICRP 1991). In the context of radon, the equivalent dose, H, is given as

$$H = D_{Low\ LET} + 20\ D_{High\ LET} \tag{4.3}$$

where $D_{Low\text{-}LET}$ and $D_{High\text{-}LET}$ denote the absorbed doses due to electrons and photons of low linear energy transfer (LET) and alpha particles of high-LET, respectively, and 20 is the radiation weighting factor for alpha radiation (ICRP 1991). Equivalent dose is a dosimetric quantity of radiation protection and is of limited utility in health risk assessments because the radiation weighting factor embodies consideration of the relative biologic insults of the different kinds of radiation.

For reference purposes, table 4.5a gives the equivalent dose received by various tissues of adults, assuming an intake of a unit activity of ^{222}Rn dissolved in water. Results are presented for the base case and the bounding cases describing the extent to which radon is assumed to diffuse into the blood vessels and tissue of the stomach wall. The dose to the stomach calculated for these cases can be compared with the values in table 4.4, which were extracted from the literature.

The equivalent doses received by individuals of various ages, assuming an intake of a unit activity of ^{222}Rn in water, is given in table 4.5b for the base-case assumption regarding radon uptake in the stomach wall. These dose values reflect

TABLE 4.5a   Committed Equivalent Dose per Unit Activity of ^{222}Rn Ingested (Sv Bq^{-1}) in the Adult as a Function of Diffusion Into the Stomach Wall

Organ	Uptake in Stomach Wall		
	No Diffusion	Base Case	Saturated Diffusion
Adrenals	7.8E-10	2.0E-10	3.0E-10
Bladder	3.4E-10	9.9E-11	1.4E-10
Endosteal Tissue	7.0E-09	1.8E-09	2.8E-09
Brain	7.7E-10	2.0E-10	3.0E-10
Breast Tissue	2.8E-10	8.5E-11	1.2E-10
Stomach Wall	8.9E-10	2.4E-08	3.1E-07
Small Intestine	1.2E-09	1.6E-10	3.1E-09
Upper Large Intestine	1.5E-09	1.3E-10	5.5E-10
Lower Large Intestine	2.6E-09	1.7E-10	6.5E-10
Kidneys	4.2E-09	1.2E-09	2.0E-09
Liver	1.1E-09	1.7E-09	2.1E-09
Muscle	5.0E-10	1.4E-10	2.0E-10
Ovaries	3.1E-10	8.7E-11	1.3E-10
Pancreas	8.8E-10	9.3E-11	3.4E-10
Red Marrow	7.2E-09	1.8E-09	2.7E-09
Spleen	6.8E-10	1.4E-10	3.0E-10
Testes	5.8E-10	1.5E-10	2.2E-10
Esophagus	2.8E-10	8.4E-11	1.2E-10
Thyroid	7.6E-10	2.0E-10	2.9E-10
Uterus	3.0E-10	8.6E-11	1.3E-10
Lung	4.7E-10	1.36E-10	1.9E-10
Effective[a]	2.1E-09	3.5E-09	3.8E-08

[a]Sum of weighted equivalent doses (see ICRP 1991).

TABLE 4.5b   Committed Equivalent Dose per Unit Activity of ^{222}Rn Ingested (Sv Bq^{-1}) for Various Subjects (Base Case Assumption Regarding Diffusion of Radon into the Stomach Wall)

Organ	Age at Intake (yr)					
	Infant	1-yr	5-yr	10-yr	15-yr	Adults
Adrenals	2.5E-09	1.0E-09	5.1E-10	3.0E-10	2.3E-10	2.0E-10
Bladder	6.3E-10	4.6E-10	2.6E-10	1.4E-10	1.1E-10	9.9E-11
Endosteal Tissue	1.5E-08	1.1E-08	5.0E-09	3.3E-09	2.7E-09	1.8E-09
Brain	1.2E-09	9.9E-10	5.0E-10	2.9E-10	2.2E-10	2.0E-10
Breast	5.9E-10	4.3E-10	2.4E-10	1.3E-10	9.7E-11	8.5E-11
Stomach Wall	3.0E-07	1.6E-07	7.3E-08	4.2E-08	3.1E-08	2.4E-08
Small Intestine	1.2E-09	7.9E-10	4.2E-10	2.5E-10	1.8E-10	1.6E-10
Upper Large Intestine	9.8E-10	6.7E-10	3.8E-10	2.2E-10	1.5E-10	1.3E-10
Lower Large Intestine	1.3E-09	8.7E-10	4.9E-10	2.8E-10	1.9E-10	1.7E-10
Kidneys	6.1E-09	4.1E-09	2.1E-09	1.3E-09	9.3E-10	1.2E-09
Liver	1.5E-08	1.2E-08	3.7E-09	2.4E-09	1.4E-09	1.7E-09
Muscle	8.2E-10	6.6E-10	3.4E-10	1.9E-10	1.5E-10	1.4E-10
Ovaries	6.1E-10	4.4E-10	2.5E-10	1.4E-10	1.0E-10	8.7E-11
Pancreas	7.0E-10	4.4E-10	2.5E-10	1.5E-10	1.1E-10	9.3E-11
Red Marrow	9.5E-09	8.5E-09	4.2E-09	2.5E-09	2.0E-09	1.8E-09
Spleen	1.3E-09	4.5E-10	2.8E-10	1.8E-09	1.7E-10	1.4E-10
Testes	2.2E-09	7.5E-10	3.9E-10	2.2E-10	1.7E-10	1.5E-10
Thymus	5.9E-10	4.2E-10	2.4E-10	1.3E-10	9.7E-11	8.4E-11
Thyroid	1.5E-09	9.7E-10	5.0E-10	2.9E-10	2.2E-10	2.0E-10
Uterus	6.0E-10	4.3E-10	2.56E-10	1.3E-10	9.9E-11	8.6E-11
Lung	9.1E-10	7.2E-10	3.8E-10	2.1E-10	1.6E-10	1.3E-10
Effective[a]	4.0E-08	2.3E-08	1.0E-08	5.9E-09	4.2E-09	3.5E-09

[a]Sum of weighted equivalent doses (see ICRP 1991).

the age dependence in both the sizes of body organs and the behavior of radon and its decay products in the body. A decrease in the dose per unit intake with increasing age at intake is evident in table 4.5b. The lower consumption of tapwater during childhood results in an intake of dissolved radon during childhood that is a small fraction of the lifetime intake. Thus, despite the higher dose per unit intake at these ages, relative to that of the adult, the lower consumption rates result in intakes in the first 10 years that contribute about 30% to the lifetime risk.

CANCER RISK PER UNIT ^{222}RN CONCENTRATION IN DRINKING WATER

Estimates of the cancer mortality risk per unit concentration of ^{222}Rn in drinking water were derived with the method of *Federal Guidance Report 13* (EPA 1998). That method yields a risk estimate that applies to an average member of the public, in the sense that the estimate is averaged over the age and sex distributions of a hypothetical closed "stationary" population whose survival functions and cancer mortality rates are based on recent data for the United States. Specifically, the total mortality rates in this population are defined by the 1989-1991 US decennial life tables (1989-91; 1997), and cancer mortality rates are defined by US cancer mortality data for the same period (NCHS 1993a). The hypothetical population is referred to as "stationary" because the sex-specific birth rates and survival functions are assumed to be invariant over time.

A schematic of the method of computation is shown in figure 4.8. The main steps in the computation are shown in the numbered boxes in the figure and summarized below.

1. *Lifetime risk per unit absorbed dose at each age*

For each of 14 cancer sites in the body, radiation-risk models are used to calculate sex-specific values for the lifetime risk *per unit absorbed dose* for each

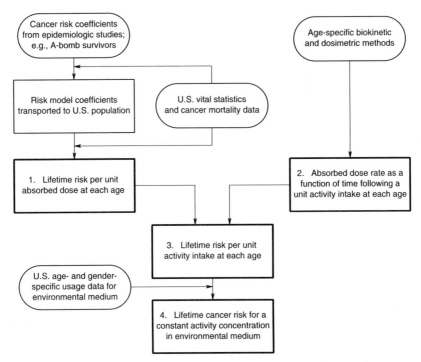

FIGURE 4.8 Schematic of method to estimate cancer mortality risk per unit concentration of Rn-222 in drinking water, derived using the methodology of *Federal Guidance Report 13* (EPA 1998).

type of radiation received at each age. These models provide a method for calculation of radiogenic-cancer risks based on a critical review of data on the Japanese atomic-bomb survivors and other study groups and methods of applying radiation risk estimates across populations.

The cancer sites considered are the esophagus, stomach, colon, liver, lung, bone, skin, breast, ovary, bladder, kidney, thyroid, red marrow (leukemia), and residual (all remaining cancer sites combined). An absolute-risk model is applied to bone, skin, and thyroid; that is, it is assumed for these sites that the radiogenic cancer risk is independent of the baseline cancer mortality rate (cancer mortality death rate for a given site in an unexposed population). For the other cancer sites, a relative-risk model is used; it is assumed that the likelihood of a radiogenic cancer is proportional to its baseline cancer mortality rate. The baseline cancer mortality rates are calculated from US cancer mortality data for 1989-1991 (NCHS 1993a; 1993b; 1992).

The computation of sex- and site-specific values for the lifetime cancer risk per unit absorbed dose involves an integration over age, beginning at the age at which the dose is received, of the product of the age-specific risk-model coefficient (times the baseline mortality of the cancer in the case of a relative-risk model) and the survival function. The survival function is used to account for the possibility that the exposed person will die of a competing cause before a radiogenic cancer is expressed.

Estimates of the site-specific cancer mortality for a hypothetical low dose, low dose rate, uniform irradiation of the whole body by low-LET and high-LET radiations are given in table 4.6. Some organs such as male breast or brain are not explicitly included as these sites have not shown definitive dose responses (Pierce and others 1996). To the extent that these sites contribute to the total cancer response, they are included in the Residual category.

The risk estimates of table 4.6 are based on the risk-model coefficients in *Federal Guidance Report 13* (EPA 1998). The estimates are age-averaged values for the hypothetical stationary population. For details regarding the method of computation, see *Federal Guidance Report 13*.

2. *Absorbed-dose rates as a function of time after an acute intake*

Age-specific biokinetic models for radon and its decay products are used to calculate the time-dependent inventories of activity in various regions of the body after acute intake of a unit activity of ^{222}Rn. This calculation is performed for each of six "basic" ages at intake: infancy (100 d); 1, 5, 10, and 15 y; and maturity (20 y). The biokinetic model for radon used in the calculations was described above (see appendix A for further details). The biokinetic models for the radon decay products were taken from ICRP's recent series of documents on age-specific doses to members of the public from intake of radionuclides (ICRP 1996; ICRP 1995; ICRP 1993; ICRP 1989).

TABLE 4.6 Age-averaged Site-Specific Lifetime Cancer Mortality Risk Estimates (Deaths per Person-Gy) from Low-Dose, Low- and High-LET Uniform Irradiation of the Body

| Cancer Site | Low-LET | | High-LET | |
	Males	Females	Males	Females
Esophagus	7.30×10^{-4}	1.59×10^{-3}	1.46×10^{-2}	3.18×10^{-2}
Stomach	3.25×10^{-3}	4.86×10^{-3}	6.50×10^{-2}	9.72×10^{-2}
Colon	8.38×10^{-3}	1.24×10^{-2}	1.68×10^{-1}	2.48×10^{-1}
Liver	1.84×10^{-3}	1.17×10^{-3}	3.68×10^{-2}	2.34×10^{-2}
Lung	7.71×10^{-3}	1.19×10^{-2}	1.54×10^{-1}	2.38×10^{-1}
Bone	9.40×10^{-5}	9.60×10^{-5}	1.88×10^{-3}	1.92×10^{-3}
Skin	9.51×10^{-5}	1.05×10^{-4}	1.90×10^{-3}	2.10×10^{-3}
Breast	0.00	9.90×10^{-3}	0.00	9.90×10^{-2}
Ovary	0.00	2.92×10^{-3}	0.00	5.84×10^{-2}
Bladder	3.28×10^{-3}	1.52×10^{-3}	6.56×10^{-2}	3.04×10^{-2}
Kidney	6.43×10^{-4}	3.92×10^{-4}	1.29×10^{-2}	7.84×10^{-3}
Thyroid	2.05×10^{-4}	4.38×10^{-4}	4.10×10^{-3}	8.76×10^{-3}
Leukemia	6.48×10^{-3}	4.71×10^{-3}	6.48×10^{-3}	4.71×10^{-3}
Residual[a]	1.35×10^{-2}	1.63×10^{-2}	2.70×10^{-1}	3.26×10^{-1}
Total	4.62×10^{-2}	6.83×10^{-2}	8.01×10^{-1}	1.18

[a]Residual is a composite of all radiogenic cancers not explicitly identified in the table.

Age-specific dosimetric models are used to convert the calculated time-dependent regional activities in the body to absorbed dose rates for both the low-LET (photons and electrons) and high-LET (alpha) radiations to radiosensitive tissues as a function of age at intake and time after the intake. Absorbed-dose rates for intake ages intermediate to the six basic ages (infancy; 1, 5, 10, and 15 y; and maturity) are determined by interpolation.

3. *Lifetime cancer risk per unit intake at each age*

For each cancer site, the sex-specific values of lifetime risk per unit absorbed dose received at each age (derived in the first step) are used to convert the calculated absorbed-dose rates to lifetime cancer risks for the case of an acute intake of one unit of activity at each age x_i. This calculation involves integration over age of the product of the absorbed-dose rate at age x for a unit intake at age x_i, the lifetime risk per unit absorbed-dose received at age x, and the value of the survival function at age x divided by the value at age x_i. The survival function is used to account for the probability that a person exposed at age x_i is still alive at age x to receive the absorbed dose. It is assumed that the radiation dose is sufficiently low that the survival function is not seriously affected by the number of radiogenic-cancer deaths at any age.

4. *Lifetime cancer risk for chronic intakes*

For purposes of computing a risk coefficient, it is assumed that the radon concentration in the drinking water remains constant and that all persons in the population are exposed to that concentration throughout their lifetimes. It is assumed that the lifetime average drinking water consumption rate is 0.6 L d^{-1}.

For each cancer site and each sex, the lifetime cancer risk for chronic intakes is obtained by integration over age x of the product of the lifetime cancer risk per unit intake at age x and the expected drinking water consumption at age x.

Except for the calculations of the time-dependent organ activities and absorbed-dose rates, each of the steps described above is performed separately for each sex and each cancer site. A total-risk coefficient is derived by first adding the risk estimates for the different cancer sites in each sex and then calculating a weighted mean of the coefficients for males and females. The weighted mean of coefficients for males and females involves the presumed sex ratio at birth, the sex-specific risk per unit intake at each age, and the sex-specific survival function at each age.

The cancer mortality risks associated with lifetime ingestion of ^{222}Rn dissolved in drinking water at a concentration of 1 Bq m^{-3} are given in table 4.7; the total average over both sexes is 1.9×10^{-9}. The uncertainty in this estimate is associated largely with the estimated dose to the stomach and with the epidemiologic data used to estimate the risk. Cancer of the stomach is a major late effect in

TABLE 4.7   Cancer Mortality Risk Associated with Lifetime Ingestion of ^{222}Rn at a Concentration of 1 Bq m^{-3} in Drinking Water[a]

Cancer Site	Cancer Mortality Risk (m^3 Bq^{-1})		
	Males	Females	Both Sexes
Esophagus	1.5E-12	3.3E-12	2.2E-12
Stomach	1.3E-09	2.0E-09	1.6E-09
Colon	4.6E-11	7.7E-11	5.9E-11
Liver	6.8E-11	4.4E-11	5.8E-11
Lung	2.6E-11	4.5E-11	3.4E-11
Bone	5.4E-12	5.7E-12	5.5E-12
Breast	—	1.0E-11	4.5E-12
Ovary	—	6.1E-12	2.6E-12
Bladder	7.8E-12	3.7E-12	6.0E-12
Kidney	1.8E-11	1.1E-11	1.5E-11
Thyroid	1.3E-12	3.1E-12	2.1E-12
Leukemia	1.9E-11	1.5E-11	1.7E-11
Residual[b]	6.9E-11	9.2E-11	7.9E-11
Total	1.5E-09	2.3E-09	1.9E-09

[a]To express risk in the conventional units (L pCi^{-1}), multiply the values by 37.
[b]The average of the absorbed dose rates to muscle, pancreas, and adrenals is applied to this group of cancers.

the survivors of the atomic bombings in Japan, where the high susceptibility appears to be related to the high baseline rate. The basic data used in the risk calculations for the present report were derived from the atomic-bomb survivors and applied to the US population with a relative-risk model in conjunction with the US stomach-cancer experience. It is noted that the high baseline stomach-cancer incidence among the Japanese and the declining incidence in the US contribute to the uncertainty in the risk estimate. An estimate of the uncertainty in the risk was derived on the basis of judgment that the absorbed dose to the stomach is probably not greater than three times the base case of table 4.4 (divide by 20 to obtain absorbed dose) and probably greater that one-fiftieth (2%) the base case. The asymmetric bounds reflect the judgment that the base case estimate is taken to be conservative; however, at this time sufficient information is not available to further refine the model and its parameter values. Similarly, it was judged that the stomach cancer mortality coefficients are probably not greater than three times the values of table 4.6 while they probably are greater than one-tenth the tabulated values. On the basis of those judgments, it is concluded that the risk posed by ingestion of water containing ^{222}Rn at 1 Bq m^{-3} probably lies between 3.8×10^{-10} and 4.4×10^{-9}, with 1.9×10^{-9} as the central value.

SPECIAL POPULATIONS AT RISK

No information is available to identify the characteristics of individuals or to suggest that any population group that might be at increased carcinogenic risk because of the presence of radon in drinking water. A number of environmental factors have been associated with stomach cancer, although the incidence of gastric cancer has been declining during the last 50 y. Stomach cancer is essentially a disease of the poor, not only in developing countries, but also in the West, where there is an inverse correlation between stomach-cancer risk and socioeconomic status. A strong link appears to exist between the ubiquitous bacterium *Helicobacter pylori* and stomach cancer, but there is no known interaction between *H. pylori* and radiation or radon (McFarlane and Munro 1997).

Sikov and others (1992) investigated the developmental toxicology of radon exposures in the rat. They did not find any teratogenic or reproductive effects in pregnant rats exposed to airborne radon at high concentrations. Radon in the maternal blood would flow to the placenta and, depending on the relative solubilities of radon in maternal and fetal blood, could be absorbed by the fetus. However, the regional blood flow to the uteroplacental unit during the period when most teratogenic effects are possible (1st trimester) is very small (Thaler and others 1990). The same exchange is possible for either inhaled or ingested radon. Thus, it appears unlikely that radon in drinking water would have substantial teratologic or reproductive effects.

5

Dosimetry of Inhaled Radon and its Associated Risk

INHALATION OF RADON AND ITS SHORT-LIVED DECAY PRODUCTS

The occurrence of bronchogenic lung cancer after inhalation of ^{222}Rn and its short-lived decay products is well established from follow-up studies of underground miners (Lubin and others 1995; 1994; UNSCEAR 1993; National Research Council 1988, 1999; NCRP 1984a). There is convincing evidence that occupational exposure has produced excess lung cancer. The bronchial airways are the location of most lung tumors (Saccomanno and others 1996). The only quantitative risk estimates available are those from underground-miner studies. The evaluation of lung-cancer risk in miners is summarized later in this chapter.

RISK POSED BY INHALATION OF ^{222}RN DECAY PRODUCTS

The assessment of risk in miners did not rely on internal dosimetry but was based on the air exposure to decay products in units of potential alpha energy concentration (PAEC) in the mines. The unit of exposure was the working level month (WLM) and was an easy measurement to make in mine air by simply taking a filtered air sample and counting total alpha particles (NCRP 1988). For domestic (i.e., residential) exposures, 1 WLM is equivalent to an indoor radon concentration of exposure at 185 Bq m^{-3} for a full year with 70% of time spent in the home (Harley and others 1991). The SI unit of exposure replacing PAEC is the joule hour per cubic meter (J hr m^{-3}); 1 WLM equals 0.0035 J h m^{-3}. However, because the original miner studies and all of the subsequent analyses and models refer to exposures in units of WLM, we have preserved that formalism in this discussion.

Excess lung cancer in underground miners was evident after exposures to radon decay products of several hundred to several thousand WLM that would equate to long-term exposure to ^{222}Rn in the home at very high concentrations. The health effects of average home concentrations are less certain, and current residential epidemiologic studies are attempting to measure the risk (as discussed later in this chapter). However, risk estimates based on these studies are not currently available.

The exposure assessment for miners was related to the potential energy concentration in air. However, it is the actual bronchial dose that confers the lung cancer risk. Therefore, it is necessary to know whether the bronchial dose in miners per unit PAEC in air in mines yields an equivalent bronchial dose per unit PAEC in homes. If the dose per unit exposure is equal in both situations then the derived miner risk estimates may be applied directly to home exposure.

The projection of risk from the mines to other environments, particularly for domestic exposures and for the entire lifetime, also makes ^{222}Rn decay product bronchial dosimetry a necessity. That requires not only physical models to evaluate the dose per unit exposure in mines relative to that in other exposure situations but also biological risk models to compare short-term with whole-life exposure. Occupationally exposed miners were exposed for 2 y to about 20 y. As described in the BEIR VI report (National Research Council 1999), there are factors that compensate for differences in exposure conditions between mines and homes, such as unattached fraction and breathing rate. Therefore, the miner risk estimates were used directly to estimate the risk from radon and its decay product exposure in domestic environments (National Research Council 1999).

LUNG DOSE FROM ^{222}RN GAS

The alpha dose delivered to target cells in bronchial epithelium arises mainly from the short-lived decay products deposited on the bronchial airway surfaces. The alpha dose from radon gas itself is smaller than that from its decay products because of the location of radon as it decays in the airway; there is a low probability that an alpha particle will interact with a cell. The decay products, however, are on the airway surfaces within about 20 to 30 μm of these target cells, and thus have a higher probability of hitting a target-cell nucleus.

The annual weighted equivalent dose from radon gas decaying in the lung has been calculated by the International Commission on Radiological Protection (ICRP 1981) and the National Council on Radiation Protection and Measurements (NCRP 1987a; 1987b) (see Table 5.1). The gas dose itself is 7 × 10^{-3} mSv y^{-1} per Bq m^{-3} (ICRP 1981) or 5 × 10^{-3} mSv y^{-1} per Bq m^{-3} (NCRP 1987a). The dose from ^{222}Rn gas is lower by about a factor of 10 compared to the bronchial dose from the decay products deposited on the airways.

TABLE 5.1   Annual Weighted Equivalent Dose[a] to the Lung from ^{222}Rn Gas Exposure

Organ	mSv y^{-1} per Bq m^{-3}	Reference
Whole lung[b]	7×10^{-3}	ICRP (1981)
Bronchial surfaces[c]	5×10^{-3}	NCRP (1975)

[a]Weighted using ICRP values of W_r and W_t.
[b]Dose calculated from ^{222}Rn solubility in tissue and radon in airways.
[c]Dose calculated to bronchial surfaces from ^{222}Rn decay in airways.

DOSE TO ORGANS OTHER THAN THE LUNG FROM INHALED ^{222}RN

Any inhaled gas, including radon, is slightly soluble in body tissues. Radon in the lung diffuses to blood and is transported to other organs, where the gas and the decay products that build up in the tissue deliver a radiation dose. Harley and others (1958) in a study of inhaled radon, determined the solubility of radon in the body. Two persons were in a controlled, relatively high-radon atmosphere for about a day. Sequential exhaled-breath samples were used to infer retention times in the five major body-compartments—lung, blood, intracellular and extracellular fluid, and adipose tissue. The data were used in the metabolic modeling of the dose to other organs from inhaled radon (Harley and Robbins 1992). The dose to organs other than the lung had been calculated previously by Jacobi and Eisfeld (1980). The dose per unit exposure for organs other than the lung are shown in table 5.2, where it can be seen that the dose to other organs is lower than the dose to the bronchial epithelium, in most cases by a factor of about 100.

^{222}RN DECAY-PRODUCT DOSE DURING SHOWERING

The most important variables in the alpha dose to cell nuclei in the bronchial airways are aerosol size distribution, breathing rate, and location of the target-cell nuclei.

The most extensive activity-weighted size distributions that have been measured in homes were reported by Hopke and others (1995a). Figure 5.1 shows the average values of the activity fractions for each of the decay products and for PAEC for the homes in which no smokers are present; figure 5.2 presents similar data on homes with smokers.

When a home is supplied by radon-bearing groundwater, the radon that is released into the air during water use becomes another source of indoor radon decay products. To evaluate the significance of this contribution to the overall radon risk, it is necessary to examine each instance of water use (such as in the kitchen, bathroom, and laundry room) and to look at both the steady-state (long-

TABLE 5.2   The Weighted Equivalent Dose to Tissues Other Than the Lung for Continuous Exposure to 1 Bq m^{-3} of ^{222}Rn in Air

Tissue	mSv y^{-1} per Bq m^{-3}	Reference
Liver[a]	5.1×10^{-5}	Jacobi and Eisfeld (1980)
Kidneys[a]	5.6×10^{-5}	Jacobi and Eisfeld (1980)
Spleen[a]	5.2×10^{-5}	Jacobi and Eisfeld (1980)
Red Bone Marrow[a]	9.6×10^{-5}	Jacobi and Eisfeld (1980)
Bone Surfaces[a]	2.5×10^{-5}	Jacobi and Eisfeld (1980)
Soft Tissue[b]	3.0×10^{-5}	Harley and Robbins (1992)
Adipose Tissue[b]	9.0×10^{-5}	Harley and Robbins (1992)
Skin[b]	50×10^{-5}	Harley and Robbins (1992)
Normal Marrow[b]	6.3×10^{-5}	Harley and Robbins (1992)
Adipose Tissue Marrow[b]	16×10^{-5}	Harley and Robbins (1992)
Bone Surfaces (Normal Marrow)[b]	1.5×10^{-5}	Harley and Robbins (1992)
Bone Surfaces (Adipose Tissue Marrow)[b]	3.0×10^{-5}	Harley and Robbins (1992)
T Lymphocytes[b,c]	0.01	Harley and Robbins (1992)
Alveolar Capillaries[b]	20×10^{-5}	Harley and Robbins (1992)

[a]Weighted using ICRP (1977) weighting factors.
[b]Weighted using ICRP (1990) values of W_r and W_t.
[c]T lymphocytes located in bronchial epithelium.

term) and the dynamic (short-term) components of the exposure. The potentially most important source of short-term exposure is the release of radon from water during showering and the subsequent inhalation of its decay products. The steady-state component has been studied in considerable detail (Fitzgerald and others 1997; Bernhardt and Hess 1996) and is described in chapter 3. Only recently have there been studies of the time-varying exposure. The exposure assessment of waterborne radon includes both its contributions to long-term average indoor radon concentrations and the short-term, perturbed conditions that exist as a result of showering.

Showering Conditions

It is necessary to provide the radon-progeny activity size distribution as a function of time during and after showering. There have been two recent studies of the increments in exposure and dose that arise from showering with radon-laden water (Fitzgerald and others 1997; Bernhardt and Hess 1996).

Exposure conditions are quite different during showering because radon is transferred from water to air with little or no direct release of decay products from the water. As a result, there might be a high local concentration of ^{222}Rn, but it takes time for the decay products to grow into equilibrium concentrations. Because the ingrowth of the activity will occur with an effective half-life of about 30 min, the highest concentrations of decay-product activity occur after the person

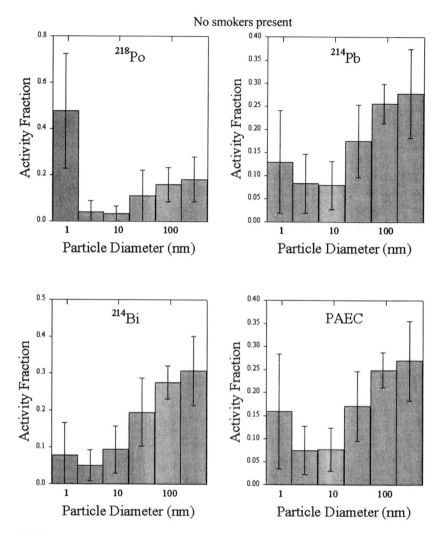

FIGURE 5.1   Average values of the activity fractions for each of the decay products and for PAEC (no smokers present). The error bars indicate the standard deviations of the 20 distributions that comprise the average.

has left the area. Thus, exposure to the [222]Rn decay products is much lower than would be expected based on radon itself.

The Fitzgerald and others (1997) study used a normally occupied dwelling that was supplied with water from a well; the water contained radon at 518,000-555,000 Bq m^{-3}. The dwelling was a one-story wooden house of a type common in northern New York state. Details of the measurement methods for radon and

Smokers present

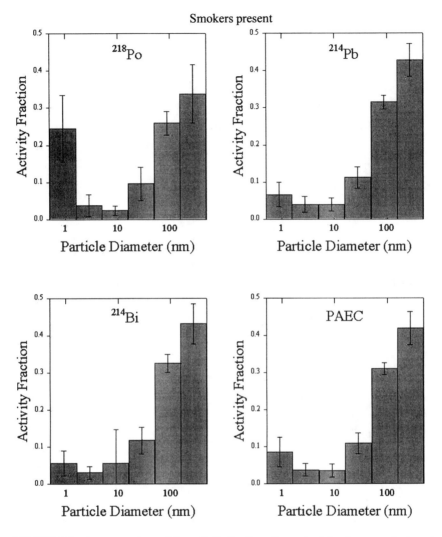

FIGURE 5.2   Average values of the activity fractions for each of the decay products and for PAEC (smokers present). The error bars are as in figure 5.1.

its decay products are provided by Fitzgerald and others (1997). Figure 5.3 shows the radon concentration as a function of time under several ventilation conditions. Figure 5.4 shows the evolution of PAEC as a function of time in one experiment.

The growth of the aerosol particles means that any calculations of dose must allow for the possibility of hygroscopic growth of the particles, such that the aerosol size spectrum and hence the activity-weighted size spectrum might shift

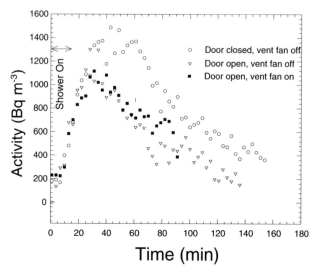

FIGURE 5.3   Radon concentration as a function of time under several ventilation conditions.

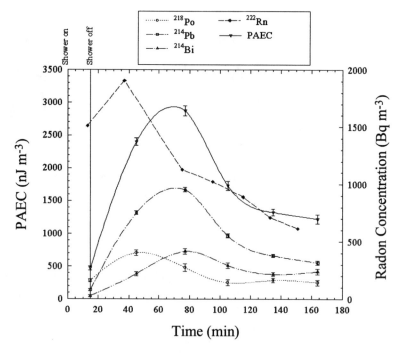

FIGURE 5.4   Evolution of PAEC as a function of time in one experiment.

size before, during, and after a shower. If an aerosol particle can grow, the increase in size will result in a larger decay-product attachment rate. If the particles can grow in the high humidities during showering, they can grow in the respiratory tract (George 1993). However, measurements show that this growth is minimal (Dua and Hopke 1996).

Changes in particle size will have two effects. First, the changes will affect the deposition of the particles onto the room surfaces, thereby affecting the amount of aerosol decay product available for inhalation. Second, the changes will alter where the aerosol particles deposit in the respiratory tract, and so affect the dose delivered to the lung (National Research Council 1991a). Furthermore, depending on how close the bathroom humidity is to 100%, the amount of possible growth when the particles enter the lungs is reduced, further altering deposition patterns in the lung. Figure 5.5 presents a series of number-weighted particle size distributions based on measurements with a scanning electrical mobility spectrometer. From the number-weighted size distributions, activity-weighted distributions can be calculated by using the attachment coefficients of Porstendörfer and others (1979). Figure 5.6 shows the estimated activity-weighted size distributions. Tu and Knutson (1991) have shown that this method provides a reasonable approximation of the directly measured activity-weighted distributions.

It is possible to make direct activity-weighted size distribution measurements (Ramamurthi and Hopke 1991; Tu and others 1991). However, the system available for such activity-weighted size distribution measurements draws 90 L

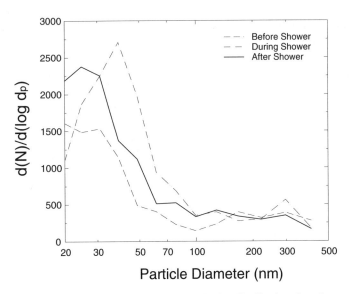

FIGURE 5.5   A series of number-weighted particle size distributions based on measurements with a scanning electrical mobility spectrometer.

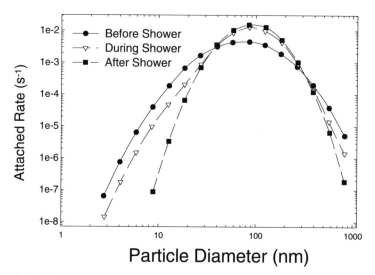

FIGURE 5.6   Calculated activity-weighted size distributions.

of air per minute and so would substantially alter the aerosol size distribution in a volume as small as a bathroom. Since it is not feasible to make direct measurements of activity-weighted size distributions, they have to be calculated. For initial work with the field data, a series of simplifying assumptions were made. From the measured number-weighted size distributions, the activity-weighted size distributions could be calculated as follows. Using the equations given by Porstendörfer and others (1979), the attachment coefficients for ^{222}Rn decay products to any size of particle can be calculated. With these coefficients and the experimental particle data, the attachment rates can be calculated; and with the steady-state equations given by Knutson (1988), the activity-weighted size distribution can be calculated (see figure 5.6). The steady-state approximation gives an upper bound to the calculated values, whereas a dynamic-model calculation (Datye and others 1997) gives results that are likely to be more representative of typical showering conditions.

Figure 5.6 clearly shows how the activity-weighted size distribution shifts during showering toward larger particles that are less efficient at delivering a dose to the bronchial tissues. Thus, although the activity suspended in the air increases because of enhanced attachment of the activity to the larger particles, the dose does not increase as sharply because the larger particles are less effectively deposited in the lung. The period during which the peak is shifted is short—around 5 to 10 min—and the particles return to their original size within about 15 min. The asymmetry in the peaks can in some measure be attributed to the variable nature of the particle size spectra over the sampling period and to the relatively poor counting statistics in the particular experiment. For dosimetric calculations, the particle size distribution during showering is therefore essentially the same as the normal house particle size distribution. This is not unexpected, because as radon gas is released from water droplets and decay products form, they interact with the normal home aerosol particles.

Integrated Measured ^{222}Rn Concentration in the Shower

The ^{222}Rn released to a shower from water was monitored in an ultra-high energy efficiency home with a private well (N. Harley, private communication). Integrated measurements spanned two 6-mo periods and indicate the average shower concentration. Data from this example home is presented.

The home monitored was in northern New Jersey and occupied by two adults and three children. The duration of all five morning showers was measured to be 60 minutes per day. The ^{222}Rn concentration was measured at the bathroom cold-water tap and in water at showering temperature that was collected from the showerhead but near the tub surface to determine ^{222}Rn loss. The long-term data are shown in table 5.3. The mean concentration in cold tapwater was 60,000 Bq m^{-3}; on the average, 6,000 Bq m^{-3} remained in the drain water after spray from the shower head.

The showerhead normally delivers water at 0.0053 m^3 min^{-1} with a mean ^{222}Rn concentration of 60,000 Bq m^{-3}. The fraction of ^{222}Rn lost (table 5.3) by the water to the air was 90% (the water concentration decreases from 60,000 Bq m^{-3} to 6,000 Bq m^{-3}) as the water fell from the showerhead to the tub floor. Thus, the shower released to the air at an average of about 290 Bq ^{222}Rn min^{-1}.

Table 5.4 shows the time-integrated ^{222}Rn concentration in the ventilator duct directly above the shower as measured with a passive, alpha-track detector. A small exhaust fan in the vent outlet operates during showering to remove water vapor from the home. The duct concentration approximates the concentration in the shower. The concentration over the two 6-mo intervals varied somewhat, as expected, because of different conditions in the shower.

From the preceding description of the aerosol particle size distribution as a result of showering, it is clear that the aerosol characteristics in showers are not substantially different from those found in the rest of the house in terms of particle number or size.

The annual dose to the bronchial airways of a person can be calculated from three factors:

- A knowledge of the ^{222}Rn concentration in the shower.
- The estimated time spent in the shower per year (about 1% for illustrative purposes).
- The central value of the ^{222}Rn bronchial dose factor, given the equilibrium factor in the shower (0.05) is 0.004 mGy yr^{-1} per Bq m^{-3}. This dose factor takes

TABLE 5.3   Measured ^{222}Rn Concentration in Cold Water and Water from Shower at Normal Showering Temperature (Shower Water Collected Near Tub Splash Surface)

Date	Cold Water (^{222}Rn Bq/L)	Shower (^{222}Rn Bq/L)	% Loss
Jun 94	67.34	29.60	
Mar 95	64.38	9.62	
Mar 95	57.72	3.55	
Apr 95	63.27	5.07	
Apr 95	56.24	1.78	
Apr 95	55.13	2.55	
May 95	67.34	2.41	
May 95	59.57	4.14	
May 95	53.28	4.81	
Jun 95	59.94	3.40	
Aug 95	64.75	10.10	
Aug 95	74.74	11.06	
Aug 95	66.97	10.58	
Aug 95	62.16	6.81	
Sept 95	46.99	3.61	
Oct 95	52.91	1.49	
Dec 95	53.65	1.55	
Jan 96	55.87	4.07	
Feb 96	57.35	4.07	
Apr 96	54.58	3.85	
Jul 96	47.73	5.14	
Average	60.00	6.00	90.00

into account the average particle size distribution observed in indoor air (Harley 1996). This is compared with the normal home equilibrium of 0.40 and a bronchial dose factor of 0.032 mGy yr^{-1} per Bq m^{-3}.

For typical home conditions, the annual shower exposure is a few percent of the whole house exposure attributed to water use. For this reason, the transfer factor is assumed to include both the whole house and showering exposure when calculating dose. The detailed factors given above permit the showering dose and whole house dose to be calculated separately for a specific case.

TABLE 5.4 ^{222}Rn Air Concentration Measured in the Shower Using Passive Alpha Track Detectors

Date	^{222}Rn (Bq m^{-3}) in Shower
Mar-Aug 95	4400
Aug 95-Jan 96	3500

LUNG-CANCER RISK POSED BY INHALATION
OF ^{222}RN DECAY PRODUCTS

There are five models for transporting lung cancer risk from the underground-mining studies to exposure in the environment: the historic NCRP (1984b; 1984a) model, the ICRP (1987) model, the model developed by the fourth National Research Council (National Research Council 1988) Committee on Biological Effects of Ionizing Radiations (BEIR IV), the National Cancer Institute model (Lubin and others 1994) derived from the pooling of 11 underground mining studies, and the model developed by the sixth National Research Council (National Research Council 1999) Committee on Biological Effects of Ionizing Radiations (BEIR VI). The BEIR VI model is similar to the NCI model, with updated data from the same 11 underground miner cohorts.

Although the ^{222}Rn decay-product exposure data are universally weak in all the miner epidemiologic followup studies, they are the only human data available from which one can derive numerical estimates of occupational and domestic lung-cancer risk. Some generalizations are possible and necessary to quantitate this risk.

NCRP Model

NCRP was the first to propose a model for environmental lung-cancer risk based on the miner data (NCRP 1984a). The model accounted for the fact that miners exposed for the first time when over 40 y old appeared to have higher lifetime risk of lung cancer than miners exposed for the first time in their 20s. The apparent lower lifetime risk for those exposed at young ages was assumed to be due to a reduction in risk with time since last exposure. Thus, earlier exposure was assumed to diminish because of cell death or repair of cells transformed by earlier exposure. The half-life for repair (or loss) was assumed to be 20 y.

One key factor noted by NCRP was that lung cancer is a rare disease before the age of 40, regardless of the population considered. Miners exposed when young did not generally appear as lung-cancer cases until the usual cancer ages were attained (50-70 y). That would account for an apparent increase in lifetime lung-cancer risk at higher ages because there would be a shorter time for transformed cells to be lost compared with the situation in persons exposed at lower ages. Miners were exposed, on the average, for less than 10 y in the Colorado, Ontario, and Czech cohorts. The total time for followup was 20 y or more, so the apparent reduction in risk with time after exposure could be observed.

The NCRP model took the form of an exponential reduction with time after exposure, with the stipulation that there was a minimal latent period of 5 y. Also, lung cancer could not appear before the age of 40. This model is known as a modified absolute-risk model. Risk is expressed after exposure without regard to other risks of lung cancer, such as smoking, but risk is modified by time since exposure.

To express lifetime risk after a single exposure, it is necessary to sum the risk over the number of years of life after exposure, taken as age of exposure to age 85. The lifetime risk, *TR*, posed by continuous exposure was expressed as the sum of lifetime risk for a single year's exposure over the total exposure interval considered.

When the model was developed, there was not enough information on the risk associated with smoking and ^{222}Rn exposure combined to separate an additional effect from this carcinogen. It was stated that the risk coefficient, *C*, could be modified when sufficient data were available.

Numerical values of lifetime risk for different models are shown in table 5.5.

TABLE 5.5   Lung Cancer Risk for Continuous Whole-Life Exposure to 4 pCi/L (148 Bq m^{-3} or 0.56 WLM/ yr at Indoor Conditions) as Predicted by Various Models of Domestic Exposure[a]

Model	Lifetime Risk, %	Model Type	Comment
NCRP (1984a)	0.50	Modified Absolute Risk. Two parameter model	Risk decreases with time since exposure
ICRP (1987)	0.90	Constant Relative Risk	
ICRP (1987)	0.62	Constant Additive Risk	
ICRP (1993b)	0.56	Single Value Risk per WLM	Adopted Lifetime Risk per WLM exposure
BEIR IV (National Research Council 1988)	1.1	Modified Relative Risk. Two time windows. Two parameter model	Risk decreases with time since exposure and decreases with very high exposures
NIH (Lubin and others 1994)	1.8 1.8	Modified Relative Risk. Three time windows, age and exposure rate. Three parameter model	Risk decreases with time since exposure and decreases with very high exposures
BEIR VI (National Research Council 1999)	2.0	Modified Relative Risk. Three time windows, age and exposure rate. Three parameter model.	Risk decreases with time since exposure and decreases with very high exposures.
Meta-analysis 8 domestic case-control (Lubin and Boice 1997)	0.7	Observed mortality	Linear regression fit to data from 8 domestic studies

[a]Exposure assumes a home concentration of 148 Bq m^{-3} (4 pCi L^{-1} or 0.56 WLM), calculated with 40% decay product equilibrium, and actual exposure is 70% of the home exposure.

ICRP Model

ICRP (1987) developed its models for environmental risk on the basis of both a constant-relative-risk model and a constant-absolute-risk model. ICRP assumed that the risk expressed over the years that cancer occurs would be increased if exposure occurred in childhood. It assumed that the risk was 3 times as great for exposure at ages 0-20 than for exposure at ages over 20. There is little justification for that assumption, as later information suggests that those exposed as children might have no different risk than those exposed as adults (Lubin and others 1995; Xuan and others 1993). This is discussed further later in this chapter.

The constant-absolute-risk or constant-relative-risk model is no longer considered appropriate for lung cancer. The best models use modifications of the parameters to account for a risk reduction with time since exposure. Although not biologically correct, risk estimates calculated with a constant-relative-risk model are within a factor of 3 of those calculated with other models.

Values of lifetime risk for the ICRP model are shown in table 5.5.

BEIR IV Model

The fourth National Research Council Committee on Biological Effects of Ionizing Radiations (BEIR IV) prepared a report, *Health Risks of Radon and other Internally Deposited Alpha-Emitters* (National Research Council 1988). The committee was given the raw data or selected parts of the original data from four mining cohorts: the US (Colorado), Canadian (Ontario and Eldorado), and Malmberget (Swedish) cohorts. Reanalysis was performed with the *AMFIT* program developed for analysis of the Japanese atomic-bomb survivor data. The program uses Poisson regression to estimate parameters.

With *AMFIT*, the data were analyzed with both internal and external cohorts for a control population. The BEIR IV committee stated that a relative-risk model fit the observed mortality well. The relative-risk model assumes that radon decay product exposure increases the age-specific lung-cancer mortality rate in the population by a constant fraction per WLM of exposure. However, in all cohorts, there was an obvious reduction in lung-cancer relative risk with time after exposure. The relative-risk model was modified to reduce risk with time since exposure. The BEIR IV committee called its modified relative-risk model a time-since-exposure (TSE) model.

Smoking was examined as a confounder. The only study with complete smoking history on the miners was the Colorado study. The effect was tested with a hybrid relative-risk model that incorporated a mixing parameter for smoking. A parameter value of zero fit an additive effect of smoking and ^{222}Rn interaction; a value of 1 fit a multiplicative model best. A maximum log-likelihood test was applied to the data, and it was found that the best parameter fit was between 0 and 1. This indicated that combined risk was more than additive but less than multiplicative. That is, the lifetime risk of lung cancer posed by radon exposure did not simply add to the lifetime risk of lung cancer associated with smoking, but neither did the risks multiply. The risk related to radon and smoking appeared to be between the two extremes.

The exposure data on the Eldorado cohort were not considered carefully by the BEIR IV committee. A reported Eldorado mining exposure of 1 WLM gave a 50% excess lung-cancer mortality—clearly an erroneous value. It is known that the miners had prior exposure in other mines, but the additional exposure is not known (Chambers and others 1990). The exaggerated risk per WLM in this study for the 1-WLM exposure cohort is important in controlling the overall BEIR IV model. This exposure category included a large number of person-years. There-fore, when the four cohorts were combined to yield a "best estimate" of the relative-risk coefficient, the 1-WLM group carried substantial weight. If this data point were omitted, the risk coefficient in the model would be less than the 0.025/WLM used in the final BEIR IV model. That possible effect should be carefully considered in any future models that use the Eldorado cohort. Thus, considering that inaccuracies might be incorporated into the BEIR IV model, the calculated risk estimates for both smokers and nonsmokers at environmental exposures are likely to be overestimates.

The values of lifetime risk as calculated by the BEIR IV TSE model are also shown in table 5.5.

NCI Model

The National Cancer Institute coordinated an effort to pool the epidemio-logic data from 11 underground-mining studies. The authors from the various countries pooled results, and these were reported by the National Institutes of Health (Lubin and others 1995; 1994). The report *Radon and Lung Cancer Risk: A Joint Analysis of 11 Underground Miners Studies*, is the most complete analy-sis of the health detriment to underground miners. For the pooled analysis, there were 2,701 lung cancer deaths among 68,000 miners who accumulated about 1.2 million person-years of exposure.

In all 11 cohorts, the excess relative risk (ERR) of lung cancer (the fractional increase in lung cancer) was linearly related to the cumulative exposure estimated in working level months (WLM). Thus, although other carcinogens might be present in the mine atmosphere, a clear exposure-response relationship was asso-ciated with ^{222}Rn decay products. Smoking history was complete only in the Colorado mining cohort. Because of the lack of smoking information the com-bined risk for smoking and radon could only be inferred qualitatively.

The Colorado uranium-miner data are shown as a typical example of the 11 cohorts in figure 5.7. The ERR/WLM for all 11 studies is shown in figure 5.8. Parts a and b of figures 5.8 show the ERR for all the cohorts combined, and for all cohorts with exposure under 600 WLM.

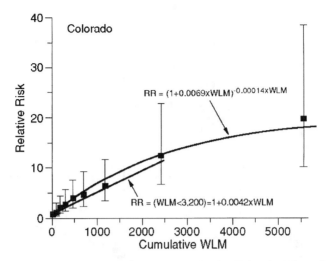

FIGURE 5.7 Relative risk (*RR*) of lung cancer in the Colorado Miner Cohort as a function of cumulative WLM exposure and fitted model. Fitted linear model shown < 3200 WLM. (NIH 1994).

One important aspect of the data is shown in figure 5.7—that the ERR at high exposures tends to flatten out. That observation is erroneously called the inverse-exposure effect. It is usually stated that the increase in lung-cancer risk per unit exposure is higher for low exposures than for high exposures. The flattening of the response curve is probably the result of cell-killing due to multiple traversals of cell nuclei. At low exposures, even a single traversal of a cell nucleus by an alpha particle is rare. Therefore, the effect is actually a reduced response at high exposure that is due to sterilization, not an increased response at low exposure.

The terminology has caused considerable confusion because it implies that domestic exposure can somehow be "more dangerous" than mine exposure. That is not the case, and it has been demonstrated that no additional risk above the linearity shown in all the cohorts is posed by domestic exposures.

The main features of the lung-cancer risk model derived from the jointly analyzed data are as follows:

1. There is a reduction in risk after cessation of working in the mines. It is called the time-since-exposure effect (the TSE factor).

2. There appears to be no clear age-at-start-of-exposure effect; that is, the age at the start of mining is not an obvious factor. However, the age attained after the start of mining is a factor, and risk decreases with age (the *AGE* factor).

3. Longer duration (the *DUR* factor) or lower ^{222}Rn concentration (the *WL*

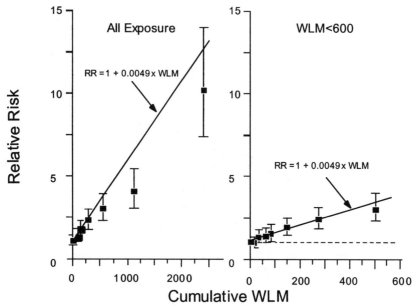

FIGURE 5.8   Fitted exposure response model (left side) and relative risk (*RR*) of lung cancer (right side) in all 11 cohorts of miners for all exposures, and for exposures < 600 WLM (Lubin and others 1995).

factor) gives rise to larger risk. Because this is how the model parameters are derived, it gives rise to the reason for the so-called inverse-exposure effect.

4. There is a higher lung-cancer risk per unit of ^{222}Rn exposure for smokers than for nonsmokers.

The two models derived from the joint analysis are considered equally likely to fit the observations.

* *TSE/AGE/WL* model (time since exposure, age, and concentration):

$$RR = 1 + \beta(w_{5\text{-}14} + \theta_2 w_{15\text{-}24} + \theta_3 w_{25}) \, \phi_{age} \, \gamma_{WL}$$

where

$w_{5\text{-}12}$, $W_{15\text{-}24}$ = exposure in WLM 5-14 years before the end of mining, and so on.

* TSE/AGE/DUR model (time since exposure, age, and duration):

$$RR = 1 + \beta(w_{5\text{-}14} + \theta_2 w_{15\text{-}24} + \theta_3 w_{25}) \, \phi_{age} \, \gamma_{DUR,}$$

The estimated parameter values derived for these two models are given in table 5.6. The combined effect of smoking and ^{222}Rn exposure could not be determined quantitatively. The pooled analysis showed a linear increase in risk of about a factor of 3 for smokers over never-smokers. This value is a decrease from the BEIR IV estimate of 10 to 1 for smokers over never-smokers. A striking feature of the data is the time-since-exposure effect. With three time windows in the model, the joint analysis showed a reduced risk with time, compared with two time windows included in the BEIR IV model, which did not show this effect.

BEIR VI Model

In 1999, the National Research Council published the BEIR VI report (National Research Council 1999), a revision and update of the BEIR IV report. The BEIR VI models to project lung-cancer incidence from ^{222}Rn decay-product exposure produce equations that are identical with that reported by NIH (Lubin

TABLE 5.6   Parameter Estimates from BEIR VI (National Research Council 1999) Models Based on Original (Lubin and Others 1994) and Updated Pooled (Lubin and Boice 1997) Miner Data

Exposure-Age-Duration Model[a]			Exposure-Age-Concentration Model[a]		
	Original Data	Updated Data		Original Data	Updated Data
$\beta^b \times 100$	0.39	0.55	$\beta \times 100$	6.11	7.68
Time-since-exposure windows					
θ_{5-14}	1.00	1.00	θ_{5-14}	1.00	1.00
θ_{15-24}	0.76	0.72	θ_{15-24}	0.81	0.78
θ_{25+}	0.31	0.44	θ_{25+}	0.40	0.51
Attained age					
$\phi_{<55}$	1.00	1.00	$\phi_{<55}$	1.00	1.00
ϕ_{55-64}	0.57	0.52	ϕ_{55-64}	0.65	0.57
ϕ_{65-74}	0.34	0.28	ϕ_{65-74}	0.38	0.29
ϕ_{75+}	0.28	0.13	ϕ_{75+}	0.22	0.09
Duration of exposure			Exposure rate (WL)		
$\gamma_{<5}$	1.00	1.00	$\gamma_{<0.5}$	1.00	1.00
γ_{5-14}	3.17	2.78	$\gamma_{0.5-1.0}$	0.51	0.49
γ_{15-24}	5.27	4.42	$\gamma_{1.0-3.0}$	0.32	0.37
γ_{25-34}	9.08	6.62	$\gamma_{3.0-5.0}$	0.27	0.32
γ_{35+}	13.6	10.2	$\gamma_{5.0-15.0}$	0.13	0.17
			$\gamma_{15.0+}$	0.10	0.11

[a]Parameters estimated on the basis of the model $RR = 1 + \beta w^* \phi_a \gamma_z$ fit using the two-stage method where $w^* = w_{5-14} + \theta_{15-24} w_{15-24} + \theta_{25+} w_{25+}$. Here the subscript a denotes categories of attained age and the subscript z denotes categories of either exposure duration (in years) or radon concentration in WL.
[b]Units are WLM^{-1}.

and others 1994). The BEIR VI analysis used updated versions of the 11 miner cohorts, so the model results are quite similar. The parameters estimated by the BEIR VI committee are also provided in table 5.6.

The NCI report summarized the calculated deaths in the US population from the assumed average exposure of 46 Bq m^{-3}. Its calculated value was 15,000 deaths, consisting of 10,000 deaths in smokers and 5,000 in non-smokers. This estimate of the attributable risk is derived from the model of lung cancer risk, a distribution of radon exposures, and the lung cancer and overall mortality rates (National Research Council 1999; NIH 1994).

EPIDEMIOLOGY OF CHILDHOOD EXPOSURE AND LUNG-CANCER RISK

It is apparent that leukemia and breast cancer are more frequent in people exposed to radiation in childhood than in those exposed as adults (National Research Council 1990a). Concern has been expressed that the same might be true for lung cancer that results from exposure to radon and its decay products at early ages.

Some data are available on occupational exposure to high radon concentrations in childhood. Lubin and others (1990) analyzed data on Chinese tin miners in the Yunnan province. Of exposed workers, 37% started employment under the age of 13; in this group, the risk coefficient for lung cancer was 1.2% WLM^{-1}. For those first employed over the age of 18, the risk coefficient was 2.9% WLM^{-1}. Later and more complete information on the entire cohort (Xuan and others 1993) showed heterogenous results with no pattern that would support the notion that children are at higher risk. It appears that children are not a particularly sensitive population.

Tentatively, the lower risk coefficient for children than for adults reported by Lubin and others (1990) suggests that the reduction in lung-cancer risk with time after exposure might be effective in children. The fact that lung cancer does not appear at a substantial rate in any population before the age of 40, permits a substantial interval for risk reduction for exposure in childhood.

ENVIRONMENTAL AND DOMESTIC EPIDEMIOLOGY

Lung-cancer excess (above that expected from smoking) in the underground-miner populations has been demonstrated conclusively. Combining that and the knowledge that some homes have radon and daughter concentrations near or above those found in historical mines, it seems virtually certain that environmental radon is responsible for some lung cancer in the general population.

There are more than 20 environmental epidemiologic studies of radon exposure to determine whether health effects can be documented directly (National Research Council 1999; Neuberger and others 1996; Neuberger 1989; DOE/CEC

1989; Borak 1988). Most domestic studies show either a slight positive or slight negative correlation between measured radon in the home and lung-cancer mortality.

Many studies are ecologic exercises that relate lung-cancer mortality in a region with indoor radon concentration (Cohen 1992; 1990). In some cases the radon is not measured, but rather is estimated as high or low, depending on the type of house. The ecologic studies are ambiguous because no attempt is made to determine actual exposure to individuals in the area of study and no correction can be made for smoking, the strongest confounder for lung cancer (Stidley and Samet 1994; Cohen 1989). The ecologic study of Cohen (1995) is the most comprehensive. It encompasses about 300,000 radon measurements in 1,601 counties in the U.S. The trend of county lung cancer mortality with increasing home radon concentration is strikingly negative, even when attempts are made to adjust for smoking prevalence, and 54 socioeconomic factors. The measured average county radon concentrations do not exceed 300 Bq m^{-3}, thus the typical low home exposure region is studied. This finding contradicts the existing risk estimates at low exposure, and a sound reason for the significant negative trend should be sought.

To date, there are 8 published case-control studies that compare the relative risk of lung cancer between high- and low-exposed groups. An attempt is made to measure the ^{222}Rn exposure in the home. The largest case-control study to date was performed in Sweden (Pershagen and others 1994). There were 1,360 cases and 2,847 controls, and exposure was assessed with 3-mo measurements during the heating season, retrospectively for homes lived in for more than 2 y after 1947 up to 3 y before diagnosis of cancer. The lung-cancer excess was not statistically significant even for smokers or nonsmokers with over 400 Bq m^{-3} in the home for over 32 y.

A meta-analysis was performed with results of the eight published domestic studies. The lung-cancer excess is not statistically significant, but the trend with increasing concentration in the homes is significant (Lubin and Boice 1997). The graph of the eight studies from Lubin and Boice (1997) is shown in figure 5.9.

All that can be said about domestic risk is that it is low and difficult, if not impossible, to detect given the high background lung-cancer mortality in the populations studied. Although a pooling of data from the largest current and past case-control studies from all countries will be performed at NCI by the year 2000, it is unlikely to provide quantitative domestic risk estimates. Because of the poor precision of the individual studies. Lung cancer from environmental exposure might eventually be documented, but it is most likely that numerical risk estimates for lung cancer from ^{222}Rn and decay-product exposure will rely on projection models from the underground-miner experience.

The difficulty in pooling the domestic studies is described by Neuberger and others (1996).

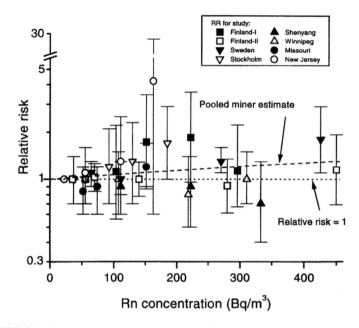

FIGURE 5.9   Relative risks (*RR*) from eight lung cancer case-control studies of indoor radon levels.   Dashed line is extrapolation of risk from miners for a 25 y exposure. Dotted line is a relative risk of 1 (Lubin and Boice 1997).

EPIDEMIOLOGY OF CANCER OF ORGANS OTHER THAN LUNG

Two followup studies of the cancer risk in organs other than the lung were performed. One cohort was uranium miners in West Bohemia, and the other was iron miners in Sweden. In the West Bohemia study of 4,320 uranium miners, 28 sites of cancer mortality were evaluated. No statistically significant cancers other than lung cancer occurred. The authors state that the possible exception is cancer of the gall bladder/extrahepatic ducts, with 12 deaths (ratio of observed to expected (O/E), 2.26; and confidence interval (CI), 1.16-3.94), but they state that such cancers would have to be studied further to prove that radon was causal (Tomasek and others 1993).

In the Swedish study, the mortality of 1,415 iron miners was studied, and 27 sites of cancer were evaluated Darby (1995). There were no statistically significant increases in gallbladder or extrahepatic duct cancers, but a marginal excess of stomach cancers was found (O/E=1.45; CI=1.04-1.98). The authors state that the stomach cancer was probably due to the considerable number of Finns in the workforce; stomach cancer in males, especially in northern Finland, is considerably higher than in Sweden, the population used for reference.

Stomach cancer in the United States has been qualitatively linked to smoking habits in some studies though no analytic study has yet been mounted to attempt to derive risk estimates. Data in the United States (NIH 1996) generally show an increasing mortality risk ratio for stomach cancer with increasing smoking rates, but the data are inconclusive. The two studies of other cancers provide reasonable support for the conclusion that the dosimetric significance of inhaled ^{222}Rn and decay products for the induction of cancer in other organs is absent or minimal.

EVALUATION OF RISK PER UNIT EXPOSURE FROM INHALED ^{222}RN IN AIR

The lifetime risk of lung cancer associated with indoor radon concentration of 150 Bq m^{-3}, calculated from the various models, is summarized in table 5.5. The most recent estimate of risk of lung cancer in the United States due to inhalation of radon decay products is from the model published in the BEIR VI report.

The BEIR VI calculated estimates of lung cancers per year in the United States from the two models are 15,000 and 21,000. These result from an average exposure of the population to 46 Bq m^{-3}; the exposure distribution was documented from the national residential radon survey by the Environmental Protection Agency (Marcinowski and others 1994).

The lifetime risk can be calculated per unit Bq m^{-3} directly from the relative risk tables given in BEIR IV, NCI and BEIR VI for smoking and nonsmoking males and females. The fractional values used by the BEIR VI committee for ever-smoking males and females are 0.58 and 0.42, respectively. The values for the lifetime base risk of lung cancer in ever-smoking males, never-smoking males, ever-smoking females, and never-smoking females are 0.116, 0.0091, 0.068, and 0.0059, respectively (J. Lubin, personal communication).

The lifetime inhalation risks per unit of exposure to ^{222}Rn in air and in air from water use are shown in table 5.7. The risk is estimated for ever-smokers and never-smokers and men and for women with the average of the two BEIR VI preferred risk models. The lifetime risks are derived as a product of the BEIR VI preferred relative risk estimates and the baseline lung cancer risks given above. The lifetime risk per unit of exposure to ^{222}Rn in air derived from water use is the risk in air multiplied by the average water transfer coefficient.

All the risk computations described above are based on up to 11 cohorts of underground miners. The atmospheric characteristics in the various mines have a wide array of values with regard to unattached fraction, decay product equilibrium, etc. The absorbed dose delivered to the target cells in bronchial epithelium differs somewhat among mines and among different homes. The various factors such as higher unattached fraction in homes versus mines and lower breathing rates in homes versus mines compensate, in such a way that the application of the

TABLE 5.7   Lifetime Risk of Lung Cancer for Lifetime Exposure to 1 Bq m^{-3} Calculated from the BEIR IV, NCI, and BEIR VI Lifetime Relative Risk Tables

Model	Lifetime Lung Cancer Risk per Bq m^{-3} in Air (in Water)		
	Men	Women	Population
BEIR IV (National Research Council 1988)			1.0×10^{-4} (1.0×10^{-8})
Lubin and others (1994)			1.2×10^{-4} (1.2×10^{-8})
BEIR VI (National Research Council 1999)			
Ever-smokers	$3.1 \ \times 10^{-4}$ $(3.1 \ \times 10^{-8})$	$2.0 \ \times 10^{-4}$ $(2.0 \ \times 10^{-8})$	
Never-smokers	0.59×10^{-4} (0.59×10^{-8})	0.40×10^{-4} (0.40×10^{-8})	
Population weighted average of ever-smokers and never-smokers			1.6×10^{-4} (1.6×10^{-8})

models for exposure in homes derived directly from the mines is considered valid for predictive purposes (National Research Council 1999).

The population estimate, 1.6×10^{-4}, for the lifetime risk of lung cancer for lifetime exposure in the home to 1 Bq ^{222}Rn m^{-3} in air, as derived from BEIR VI, is the value adopted by this committee. It can be seen that the BEIR VI estimate of lifetime risk is higher than that for the domestic studies (see table 5.5) which support a lower risk estimate for ^{222}Rn (table 5.7).

6

Molecular and Cellular Mechanisms of Radon-Induced Carcinogenesis

The exposure of cells to densely ionizing radiation, such as radon alpha particles, can initiate a series of molecular and cellular events that culminates in the development of lung and other cancers (Hall 1994). That flow can now be described in outline, starting with the deposition of clusters of ionizations and ending in the development of cancer (Cox 1994). Ionization leads to cellular damage, DNA breakage, accurate or inaccurate repair, apoptosis, gene mutations, chromosomal change, and genetic instability (Kronenberg 1994; Ward and others 1990; Ward 1988). Radiation-induced molecular changes result in the gain and loss of functions in critical regulatory genes, which permit cells to escape from normal controls and become invasive unregulated malignancies. The process of malignant transformation involves a series of changes that follow, at least roughly, a functional and temporal sequence by which cells gradually and progressively escape from normal tissue control and acquire independence, diversity, and invasive properties (figure 6.1). Molecular changes associated with radiation carcinogenesis have mainly been investigated after higher doses and dose rates than those experienced from background levels of radon exposure. Therefore these changes are described qualitatively and the extent to which any or all occur in tissues in which a small proportion of cells have experienced a single alpha particle track remains to be determined.

CELLS AT RISK

Inhalation of radon results in exposure of lung cells to alpha irradiation from radon progeny, which are deposited in the mucus layer and can result in exposure

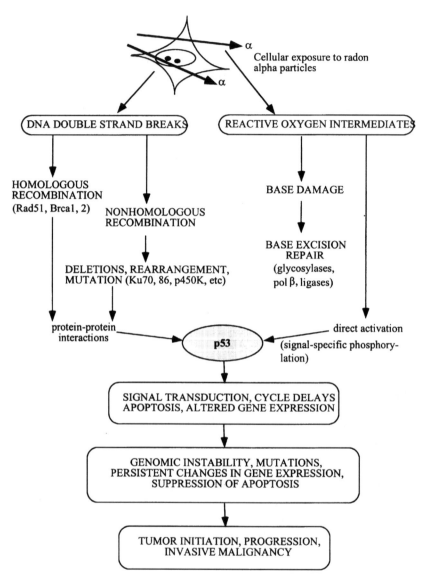

FIGURE 6.1   Flow chart showing development of malignant cells from initial α-particle damage to cells. DNA strand breaks are repaired by homologous or nonhomologous (illegitimate) double strand break rejoining, and damaged bases by base excision repair. Activation of p53 protein, initiates pathways leading to cell cycle delays and apoptosis, and surviving cells may contain gene deletions, rearrangements, amplifications, and persistant genomic instability. Mutations in oncogenes, loss of function in tumor suppressors, and loss of heterozygosity produces a heterogeneous population of cells which escapes from normal cell and tissue homeostasis to become malignant.

of epithelial cells from unilateral sources on the surface. Ingestion of waterborne radon might, on first impression, similarly expose the cells of the stomach lining. After ingestion, however, radon travels as gas molecules with high mobility through cell membranes, and cells may receive a more uniform exposure. Stem cells and other proliferating cells of the stomach are found in bands at the bases of the necks of narrow invaginations of the stomach wall that constitute the secretory glands of the stomach wall (Nomura 1996). Stem cells and other proliferative cells of the stomach are major targets of radon alpha particles, but cells of the small intestine are also potential targets. After ingestion of water, radon passes into the small intestine with a half-time of about 15-20 minutes. Radon can therefore be absorbed into the bloodstream from both the stomach wall and the small intestine. The resulting exposures to most cells of the body will then be through bloodborne radon. From that point of view, the stomach might be at greatest risk of exposure from ingested, aqueous radon. The transfer of dissolved radon from water to air and its later inhalation constitute another route by which the lungs can be at risk.

Implicit in these scenarios is the idea that the cells most likely to become malignant are the stem cells and proliferative cells that retain the capacity for continued division and can fix and express permanent genetic change. Malignant cells often retain characteristic enzymatic and cellular features of their tissue of origin, so the differentiation and specialization programs of cells might be altered but not completely abrogated by the malignant-transformation process. Alpha-particle damage to genetic material becomes fixed as permanent alterations to gene structure and expression as a result of processes that involve DNA repair, replication, and cell division. The stem cells of epithelial tissues are embedded in crypts; this renders them relatively inaccessible to direct contact with ingested or inhaled radon. Stem cells will, however, still be exposed to alpha irradiation from the lumen or blood stream, from intercellular and intracellular water, and after inhalation from decay products that plate out and act as additional sources of radiation damage. An additional factor to be considered is the potential role of chronic stomach infections. A large fraction of the normal human population carry *Helicobacter pylori* infections in the stomach that can cause gastritis and, in severe cases, ulcers. The inflammation and proliferation associated with these infections can be a factor in the induction and progression of stomach cancer and have been regarded as risk factors (McFarlane and Munro 1997).

CELLULAR DAMAGE INDUCED BY RADON ALPHA PARTICLES

Alpha particles create dense ionization that leaves tracks of ion-pair clusters across cells and tissues. Cells that suffer an alpha-particle track through the nucleus are severely injured. At the low exposure conditions under consideration from waterborne and airborne radon in the home, however, less than 1% of the cells in the bronchial epithelium would experience an alpha-particle track per year. For comparison, it requires an exposure of 100 WLM to reach the level at which the average exposure to stem cells reaches one alpha particle per nucleus (Harley and others 1996). Therefore, complex considerations of dose rates and total doses that are important for miners or other people with high occupational exposure are unimportant in consideration of domestic exposure (Brenner and others 1995; Brenner 1992) (see also BEIR VI report National Research Council 1999).

Alpha particles traverse a cell in less than 10^{-12} seconds and deposit energy corresponding to about 10-50 cGy (Jostes 1996). As the particles slow down, they deposit increasing amounts of energy (linear energy transfer, or LET) per unit length of track, reaching a maximum at the end of their track at what is known as the Bragg peak. The relative biologic effectiveness (RBE) of an alpha particle is therefore variable along its track according to whether the LET reaches a maximum at the Bragg peak (Brenner and others 1995). The average track through a spherical cell nucleus can cross many individual strands of DNA, depositing energy in the form of clusters of ionizations, and produce corresponding numbers of double-strand breaks. These breaks have a complex chemistry and have been described as multiply locally damaged sites (MLDSs) (Ward 1990). Because of the track structure and the tightly coiled nature of DNA in the nucleus, there is likely to be a nonuniform distribution of DNA breaks with an excess of small fragments which might get lost or incorrectly positioned in the process of rejoining (Ritter and others 1977).

Ion clusters can also produce reactive oxygen intermediates which can damage individual DNA bases, and at high doses, alter intracellular signal transduction, reduce macromolecular synthesis, and trigger processes that resemble those from inflammatory cytokines involved in other kinds of tissue injury. A series of early experiments in the 1950 and 1960s used collimated beams of alpha particles and other kinds of radiation and demonstrated the relative importance of nuclear, cytoplasmic, and extracellular irradiation (Munro 1970b; 1970a; Smith 1964). Those experiments showed that nuclear damage was potentially lethal; nonnuclear damage could also produce detectable effects, such as reduced DNA synthesis, but it was not lethal. Extracellular damage involved reactive oxygen intermediates that could be prevented by catalase (which degrades hydrogen peroxide) (Dendy and others 1967). More recent and technically sophisticated experiments in which the effects of single alpha particles can be estimated or observed have resulted in essentially similar conclusions (Hei and others 1997; Hickman and others 1994).

Lethality of Alpha-Particle Tracks Through Cells and Tissues

The dose required to produce an average of one lethal hit to a cell (the D_{37}) corresponds to about 1.2-1.5 alpha particles per spherical nucleus (Jostes 1996). Flattened cells can withstand more tracks (up to 15 or even more), each of which crosses shorter distances through the nucleus. Lethality can be related to the net absorption of a particular amount of total energy per cell, measured along a total path length through the nucleus—either a single track through a spherical nucleus or several shorter tracks through a flat nucleus. Calculations indicate a constant probability of 0.03-0.08 for a lethal event per micrometer of track (Jostes 1996). All radon alpha-particle effects at the low doses associated with environmental exposure from water occur from the passage of single particles through a small proportion of the cells in a tissue, so the dose-effect relationship will be a linear function of dose, with no dose-rate effects. This is true because variations in exposure change the number of cells hit by an alpha particle, rather than the amount of damage per cell. To calculate cancer risk it is then necessary to know the probability that a hit cell will undergo transformation, and the latent period and its age distribution before transformation to malignancy is complete. The latent period for single cells exposed to single tracks of alpha particles is unknown, but if it were long compared to the lifespan of the individual, the cancer risk would be correspondingly reduced, as suggested by Raabe (1987).

The important cellular subpopulation for carcinogenesis is not that of the rare cells killed by alpha-particle damage, but that of the cells that survive either with direct damage to their genetic material or with altered genomic stability. Because the calculated D_{37} is more than one alpha particle per cell in very low exposures, such as to ambient air or water, most exposed cells should survive, because it is extremely rare for any cell to be hit more than once. That might also account for the strong synergism displayed between radon exposure and cigarette-smoking: initial radon exposure leaves a viable, damaged cell, which is then stimulated further by the carcinogens found in cigarette smoke (Moolgavkar and others 1993; Brenner and Ward 1992).

Low-dose exposure also raises the question of whether radon alpha particles can give rise to radiation hormesis—the phenomenon whereby very low radiation doses are stimulatory and beneficial (Ueno and others 1996). If hormesis occurs through a stimulation of some kind of repair, the low stimulating dose must induce an excess repair capacity that can mend not only the damage caused by the initial dose, but also pre-existing endogenous cellular damage. That has been observed for repair of mitochondrial oxidative damage (Driggers and others 1996) but, evidence generally is indirect and difficult to obtain. Evidence of radiation hormesis is consequently controversial and will not be further considered here. Although extranuclear damage and extracellular ionization might play a role in some biologic effects (known as bystander effects), they are unlikely to play an important role in cell-killing (Hickman and others 1994; Dendy and others 1967). The flow of events that follow the production of DNA damage and other forms of cellular damage is therefore critical in understanding the development of malignancies.

TRANSFORMATION OF CELLS BY ALPHA PARTICLES IN VITRO

Low doses of alpha particles which simulate radon exposure have been used to achieve malignant transformation of cultured cells in studies aimed at measuring their biological effectiveness and estimating carcinogenic hazards. In general, normal diploid cells, with the exception of some hamster embryo cells, have extremely low transformation rates after irradiation. Studies of transformation therefore often use cells such as mouse 3T3 in which genetic changes have already occurred that increase their overall genetic instability and hence their transformability. Although many of these studies generated linear dose-response curves over the dose ranges used (Miller and others 1996; 1995; Brenner and others 1995; Ling and others 1994), some indicated a nonlinear response with greater effectiveness at the lowest doses (Martin and others 1995; Bettega and others 1992). Considerable uncertainty, therefore, still exists about the precise shape of the dose-response relationship for transformation of cells in culture, and by implication, also for carcinogenesis. The results in general do not permit a definitive answer to be obtained for the shape of the dose-response curve at the lowest doses and dose rates, but at the same time there is no compelling evidence to adopt any one particular non-linear dose-effect relationship. The many and varied biological changes over long time periods that are involved in carcinogenesis, which are discussed in the following sections, indicate that many factors can be expected to influence the shape of the dose-response relationship.

DNA DAMAGE AND ITS REPAIR—THE CARETAKER GENES

The gene products responsible for sensing damaged DNA and carrying out repair, euphemistically called the cellular caretakers (Kinzler and Vogelstein 1997), involve a number of enzymatic systems capable of mending single- and double-strand breaks in DNA and excising damaged and mismatched bases. Double-strand breaks are the most important kinds of damage resulting from radon alpha particles. They can be repaired through at least two pathways: homologous recombination (figure 6.2) or nonhomologous recombination (figure 6.3) (Sargent and others 1997). Repair through homologous or nonhomologous recombination involves complex sets of enzymes, which share components with enzymes and gene products associated with the generation of immunoglobulin diversity, such as RAG1 and RAG2 (Melek and others 1998) and with mitotic and meiotic recombination (Jeggo and others 1995; Jeggo 1990).

Most mammalian somatic cells are in the prereplicative, G_1, phase of the cell cycle and double-strand break repair appears to involve the nonhomologous, or illegitimate, end-joining reactions (Jeggo and others 1995). In large part, that is because the homologous chromosomes in a diploid G_1 nucleus are widely separated, so nonhomologous recombination can occur at about 10^4 times the efficiency of homologous recombination (Godwin and others 1994; Benjamin and

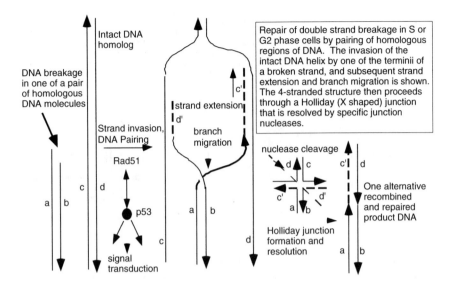

FIGURE 6.2 Mechanism of double strand break repair by homologous recombination through hybridization of the broken DNA strands sequences on the undamaged homolog. A DNA terminus is paired with the intact DNA by the action of pairing proteins including Rad51 and many other associated proteins that modulate its functions and carry out the numerous steps of pairing, elongation of DNA termini, and migration of hybridizing junction regions. Conformation changes produce a Holliday junction (a + form, 4-stranded junction) which is a strong binding site for p53, and which is resolved into separate DNA double helices containing regions of exchange, by junction-specific nucleases. The extent of sequence overlap can be very long, up to kilobases in length, and requires exact matching of DNA along most of the length of the hybrid molecules. Rad51-dependent DNA pairing is suppressed by p53-rad51 interaction, which is also a route for initiating intracellular p53-dependent signal transduction pathways. Broken double stranded DNA indicated by a,b; recipient intact strands by c,d; strands created by strand extension c', d'. De novo synthesis indicated by – – – –. Repair of a double strand break will require two of these homologous exchange events, one for each terminus. Some resolved DNA products may be visualized at the chromosomal level in mitotic cells as a sister chromatid exchange.

Little 1992). The relative importance of those pathways can vary with cell-cycle stage, tissue type, developmental stage and species. Direct measurement of DNA breakage and repair indicates that double-strand breaks can be rejoined rapidly—within a few hours. There is, however, a residuum of unrepaired damage that is greater for densely ionizing radiation, such as alpha particles, than for x rays (Ager and others 1991; Iliakis 1991; Iliakis and others 1990; Ward 1990). Although it is unknown if high levels of alpha-particle damage saturate DNA-repair systems, such potential saturation would not be relevant at low-dose ambient

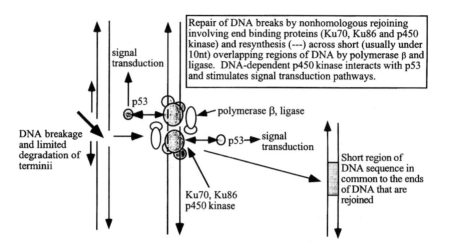

FIGURE 6.3   Mechanism of nonhomologous (illegitimate) recombination at sites of double strand breakage in DNA. The ends are sites of association of end-binding proteins, Ku70, Ku86, and p450 DNA-dependent kinase. After limited exonucleolytic degradation, short single stranded DNA termini (that may not necessarily be from either side of the original break) with a few nucleotides that can form base-pairs will hybridize and local regions can then be patched by DNA polymerase β and ligase. The extent of sequence overlap is very short, usually less than 10 nucleotides (more often 1 to about 5). The p450 kinase interacts with p53 and initiates intracellular signal transduction pathways.

radon exposure in which all damaged cells receive single alpha-particle tracks and the maximal track length through the nucleus is still below the D_{37} dose required for cell-killing. More important, even at low doses, is whether any kinds of damage are completely irreparable and whether repair is always accurate. Persistent genetic changes caused by radiation must then be caused by repair that misassembles broken termini from distant regions of the genome and triggers a lasting genetic instability.

Homologous rejoining involves matching of a broken fragment with the corresponding region on the undamaged homologous chromosome followed by strand invasion and reconstruction of the damaged region by replication of the sequence information in the intact homologue (figure 6.2). It requires that the two homologues are within range of each other; consequently, it might be more important for replicating cells in late S and G_2 phases of the cell cycle when sister chromatids are in close apposition (Takata and others in press; Sonoda and others 1998; Thompson 1996) and contributes to increased radio-resistance in these phases of the cell cycle (Cheong and others 1994). This form of double-strand repair is likely to be highly accurate because of the use of sequence information from the intact chromosomal homologue (chromatid) in reconstructing the broken DNA. The Rad51 protein plays a major role in carrying out the initial pairing reaction during homologous recombination, and dense complexes can be detected in the nuclei of irradiated cells (Haaf and others 1995), which are thought to be part of the homologous rejoining complexes (Scully and others 1997b; Scully and others 1997a). Rad51 is inhibited by association with the tumor suppressor p53 (Buchhop and others 1997) and interacts with the breast-cancer-specific gene products Brca1 and Brca2 (Scully and others 1997b; Scully 1997a). Knockout of the Rad51 and the Brca1 and Brca2 genes result in early embryo death (Lim and Hasty 1996; Tsuzuki and others 1996); this suggests a complex regulatory scheme for homologous recombination during development and carcinogenesis.

The nonhomologous recombination pathway for repair of radiation-induced DNA breakage in somatic cells involves an end-to-end rejoining reaction in which broken ends of DNA are braced by a set of supporting proteins. The gap between DNA ends is bridged by overlapping single-strand termini that are usually less than 10 nucleotides long (more commonly one to five long) and a set of proteins, including Ku70, and Ku86, p450 kinase, and DNA ligase IV (Kirchgessner and others 1995; Lees-Miller and others 1995; Getts and Stamato 1994; Rathmell and Chu 1994; Smider and others 1994; Taccioli and others 1994; Anderson 1993) (figure 6.3). The p450 kinase interacts with p53, the major signaling protein that regulates cell-cycle control, apoptosis, and the transcription of many downstream genes (Elledge and Lee 1995; Kastan and others 1995; Lane 1993). Defects in p450 have been associated with the systemic combined immunodeficiency (scid) phenotype in mice (Kirchgessner and others 1995). Knockout of the Ku70 and Ku86 genes renders cells more sensitive to ionizing radiation but, unlike the genes involved in homologous recombination, does not result in embryo death.

The rejoining reaction results in a junction made by an overlap of a few bases at each terminus with additional possibilities of single-base or larger insertions, deletions, or mismatches. No consistent DNA-sequence motifs have been found in these short regions of sequence overlaps, despite direct investigation of microsatellite repeats and telomere and triplet repeat sequences. Insertions can be many kilobases long and can come from locally produced fragments or from single-strand invasion into proximal regions of DNA. The ends involved in rejoining reactions are not necessarily those from either side of the initial break but can be from other breaks made by the same alpha track. The intervening stretch of DNA can then be lost, with consequent chromosomal rearrangement. These losses and rearrangements can involve many kilobases of DNA, producing the losses, deletions, and rearrangements of genetic material which are hallmarks of genetic effects caused by densely ionizing radiation (Zhu and others 1996; Kronenberg and others 1995; Nelson and others 1994; Phillips and Morgan 1994). The process of DNA breakage and rejoining therefore initiates a major change in signal transduction and cellular regulation that can persist over many cell generations (see discussion of genetic instability below).

The ion pairs that do not directly damage DNA can produce reactive oxygen intermediates. These intermediates influence the stability of p53 with downstream effects on cell regulation and can activate many cellular systems that are sensitive to the redox state of the cell, such as the fos/jun transcriptional regulators (Xanthoudakis and others 1996). Reactive oxygen intermediates can produce oxidative damage, of which 8-oxy-guanine is a major product. Oxidations are produced in DNA and in both deoxyribose and ribose triphosphates. Oxidized nucleotides can be incorporated into DNA and RNA, and lead to either DNA mutations or transcription and translation errors. Oxidized nucleotides can be eliminated from the nucleotide pool by MutT, which hydrolyzes 8-oxo-dGTP and 8-oxo-rGTP to monophosphates, thereby removing the oxidized bases from the pool of DNA and RNA precursors (Taddei and others 1997). MutT activity reduces mutations from naturally occurring oxidative reactions by a factor of about 10^4.

Oxidative damage involves production of damaged individual bases, such as 8-oxy-G, and many other products in DNA that cause point mutations by mispairing during DNA replication (Singer 1996) and that are repaired by the base-excision repair system. Base excision involves a set of glycosylases with limited ranges of substrate specificity (uracil, 3-methyladenine, formamidopyrimidine, glycosylases and others). The glycosylases remove damaged bases (Cunningham 1997; Singer and Hang 1997), leaving apurinic sites that are later cleaved by apurinic endonuclease (Hang and others 1996), and the gaps are replaced by polymerase β and completed by ligase I or III (Sancar and Sancar 1988). Several base-excision repair enzymes have multiple additional functions: the AP endonuclease is also known as Ref-1 and reduces the oxidized transcriptional regulators fos/jun (Xanthoudakis and others 1996); and pol β and ligase III are linked by structural protein XRCC1, which interacts with poly (ADP-ribose) polymerase (PARP) (Caldecott and others 1994). PARP is a major chromatin protein that is activated by DNA strand breaks and can exhaust the cellular NAD content by polymerization and hydrolysis (Cleaver and Morgan 1991).

DNA breaks and other base damage therefore are the assembly points for complex, multifunctional, multipurpose structures that signal their presence to many other cellular processes and within which repair and genetic changes occur. The combined actions of these cellular caretakers produces surviving cells that bear the permanent marks of alpha particle exposure, including deletions, insertions, amplifications, point mutations, and altered cellular regulation (Kronenberg and others 1995; Kronenberg 1994).

DELETION MUTAGENESIS AND CHROMOSOMAL CHANGES CAUSED BY DENSELY IONIZING RADIATION

The end result of DNA breakage and rejoining is the deletion, insertion or rearrangement of various amounts of genetic material, from a few base pairs through many kilobases to cytogenetically visible chromosomal changes (Sankaranarayanan 1991). Chromosomal fragments that are not rejoined can be excluded from interphase nuclei and can form micronuclei. These micronuclei, which encapsulate p53 (Unger and others 1994) can be scored as a quantitative measure of chromosomal damage in somatic and cultured cells. The size of deletions that persist in surviving cells is determined by the initial spacing of DNA double-strand breaks and by the presence of vital genes in the intervening sequences. Deletion sizes associated with loss of function of the adenine phosphoribosyl transferase (APRT) gene, for example, are generally smaller than those associated with loss of function of the hypoxanthine phosphoribosyl transferase (HPRT) gene because of the presence of vital genes closer to APRT than HPRT (Park and others 1995; Nelson and others 1994; Fuscoe and others 1992; Morgan and others 1990; Thompson and Fong 1980). Deletion sizes and junction positions are markedly nonrandom in both the chromosomal HPRT gene and in episomal vectors that carry reporter genes. The positions of DNA breaks and the efficiency and precision of their repair are therefore strongly influenced by chromatin structure and attachment of DNA to nucleosomal and matrix proteins and the functions of flanking genes. In an experimental cell-culture system in which a single human chromosome bearing a marker gene is carried in a hamster cell line (the A_L cell line), very few of the human genes are required for cell survival, and alpha-particle damage can produce very large deletions that involve most of the chromosome (Hei and others 1997; Ueno and others 1996). This situation cannot apply to most chromosomes in a normal cell, in which deletion sizes consistent with survival will be limited by the presence of important genes distributed throughout the genome.

CONTROL OF CELLULAR RESPONSES TO DAMAGE— THE ARBITRATOR GENE

One gene product, the p53 protein (figure 6.1), plays a critical role in regulating the multitude of responses that are elicited in damaged cells, especially those involving cell-cycle arrest and apoptosis, and interacts with numerous other regulatory and repair proteins (Elledge and Lee 1995; Kastan and others 1995; Lane 1993; 1992). The p53 protein is a rapidly synthesized, but short-lived, multifunctional protein which interacts with a wide array of other cellular and viral proteins and binds to DNA in both sequence-specific and sequence-independent fashions. In the presence of damage (either DNA breaks or reactive oxygen intermediates) the lifetime of p53 increases, it is phosphorylated at specific sites that depend on the particular signal, and it acts as a transcriptional activator with downstream effects on many other genes, especially stimulating transcription of p21, which then inhibits cell-cycle progression. Alpha-particle irradiation at low exposures has been shown to result in p53 stabilization in more cells than could have experienced alpha-particle tracks: this suggests that reactive oxygen intermediates generated outside the nucleus could result in substantial changes of cell regulation (Hickman and others 1994).

The level of damage and of consequent p53 function plays a major role either in causing cell-cycle delays (through activation of the p21 gene, which blocks cells in the G_1 phase) or in initiating apoptosis. Several of the protein complexes involved in DNA breakage and repair interact with p53, including the homologous and nonhomologous recombination complexes, and the transcription-factor component of nucleotide-excision repair, so the action of repair systems leads into the signal-transduction pathways regulated by p53. Mutations in p53 are important events that occur frequently at some stage in tumor progression and are found in over 50% of all human tumors (Greenblatt and others 1994; Hollstein and others 1991). The consequent functional changes alter many facets of cellular and gene regulation. These mutational changes in p53 do not necessarily constitute the first genetic event in carcinogenesis; for example, they can occur early in sunlight-induced skin tumors (Brash and others 1991) but late in colon cancers (Kinzler and Vogelstein 1996). The p53 protein might, in fact, play a multitude of roles in cancer, from the initial response to DNA damage, through tumor initiation and progression, to final malignancy.

APOPTOSIS—THE UNDERTAKER GENES

Cells die by several routes depending on cell and tissue type and on the particular endogenous and exogenous signals experienced. Unrepaired chromosomal damage can cause "mitotic death" when cells attempt to divide; massive damage can cause necrosis, with rapid collapse of the nucleus and permeabilization of the membranes; and a regulated cell-suicide process, apoptosis, that involves activation of proteases (caspases) and nucleases that degrade the cell components in a controlled fashion can occur (figure 6.4) (Cohen 1997). Apoptosis is activated by a wide variety of complex interrelated regulatory and signal-transduction pathways initiated by specific cellular signals, by external irradiation, and by endogenous generation of oxidative products. Some of these processes may be markedly nonlinear functions of dose, since apoptosis is a tissue response which eliminates cells that have suffered more than a critical amount of damage.

Apoptosis is an important feature of normal cell and tissue function, especially when tissue remodeling is involved during embryo development, during wound healing, and after exposure to radiation or chemicals. Apoptosis involves a family of specific proteolytic enzymes (caspases) and a specific nuclease that cleaves DNA at internucleosomal sites and produces characteristic DNA fragmentation (Enari and others 1998). Apoptosis is a complex, regulated process that involves both activators and inhibitors. These molecules fine-tune a cell's response to endogenous damage, modify its redox state, and respond to its immediate environment (Enoch and Norbury 1995; Guillouf and others 1995; Kastan

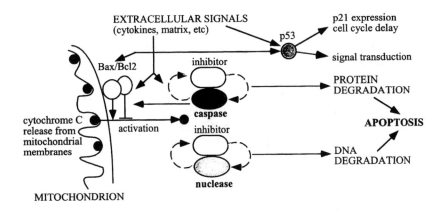

FIGURE 6.4   Mechanism of apoptosis, initiated by damage to cellular molecules or cellular signals during tissue remodeling or wound healing, resulting in cytochrome C release from mitochondria. This release is regulated by Bax/Bcl2 on the outer mitochondrial membranes, and results in activation of caspases and an apoptotic-specific nuclease. These degradative enzymes normally are associated with specific caspase-sensitive inhibitors, so that once apoptosis is set in train an autocatalytic process results with positive feedback to produce irreversible cellular degradation.

and others 1995; Kumar 1995; Leonard and others 1995; Caelles and others 1994; Canman and others 1994; Jacobson and Evan 1994; Meyn and others 1994; Lowe and others 1993; Waddick and others 1993; Uckun and others 1992). Fluctuating oxygen levels leading to oxidative bursts and the production of reactive oxygen intermediates can trigger apoptosis through their activation of p53. The mitochondria play an important role in the initial events leading to apoptosis, and one of the first signals is the release of cytochrome C into the cytoplasm (Reed 1997), which, with dATP activates a caspase cascade involving especially caspase-3 (Li and others 1997). The gene product Bcl-2 is on the outer mitochondrial membrane, where it regulates ion flow and, under conditions of normal expression, suppresses apoptosis. Its expression is induced by p53 (Pourzand and others 1997; Chen and others 1995). Two other proteins, Bax and Ced-4, bind to Bcl-2 and are inactive in bound form, but on release they further stimulate the release of apoptosis-initiating factors, which eventually activate the caspase class of proteases (Kumar 1995). One pathway to apoptosis is thus determined by the ratio of Bax/Bcl-2 expression. Many other proteins are involved in apoptosis, including many members of the caspase family of proteases and caspase inhibitors which regulate the process in different tissues and under various stresses. Other pathways by which apoptosis is activated involve the cytokines, such as TNF1-alpha, and the signal transducer and activator protein STAT1 (Kumar and others 1997) and c-myc, which interacts with Bcl-2 (Bissonnette and others 1992; Evan and others 1992; Fanidi and others 1992). The level of ATP also influences apoptosis (Eguchi and others 1997), and one of the early targets for caspases, PARP, can drain the cell of NAD and ATP in the presence of excess DNA breakage (Shah and others 1996).

Apoptosis is a normal process of cell elimination that can clear abnormal cells from the population. If apoptosis is no longer functional, abnormal cells can persist and expand in the population. The loss of apoptosis can therefore play an important role in clonal expansion during carcinogenesis. Additional genetic changes can then occur, moving a cell population through multiple stages required for the emergence of a fully malignant phenotype. Along the way, irradiated cell populations and tumor cells develop instabilities and mutator phenotypes that favor further diversity.

INITIAL GENETIC CHANGES IN CARCINOGENESIS— THE GATEKEEPER GENES

Investigation of rare cancers with strong hereditary factors—such as subsets of colon, breast, retina, and skin cancers—has suggested that genetic alterations are involved in carcinogenesis (Fearon 1997). In general, tissue-specific alterations in a small number of genes act as critical rate-limiting steps that allow cells to escape from normal controls on growth and function and to develop into autonomous populations. The genes that normally exercise these tissue-specific controls have been figuratively called gatekeepers (Kinzler and Vogelstein 1997). The autonomous cell populations acquire functions that include the suppression of apoptosis, independence from extracellular matrix signals, invasive behavior, genomic instability, the activation of oncogenes, and the inactivation of tumor suppressors. Oncogenes, such as ras, are genes that are activated by mutation or translocation and act as dominantly acting genes that induce malignant properties. Tumor suppressors, such as Rb, are genes that maintain normal cell and tissue homeostasis and whose loss permits unregulated cell proliferation to begin.

The ordered progression of genetic changes involved in carcinogenesis is most clearly understood in colon carcinogenesis (Kinzler and Vogelstein 1996). One of the earliest genes to be affected is the APC gene; it is followed by changes in DCC, p53, ras, and others. Error generation in this process is enhanced by mutations in mismatch-repair genes (Arnheim and Shibata 1997). In colon carcinogenesis, APC mutation seems to be required early, before ras mutation; when ras mutations are observed first in colon polyps, the growths are usually benign and regress; when APC mutations occur before ras mutations the tissue usually progresses to other changes that result in malignancy (Jen and others 1994). It is unlikely that such an ordered sequence of mutations or gene loss will be as clear and precise in most tissues, but the principle of an approximate sequential order to the earlier genetic changes involved in carcinogenesis seems reasonable.

The APC gene can therefore be regarded as an example of a rate-limiting gatekeeper gene that presents an initial barrier to be overcome in the initiation and progression of colon cancer (Kinzler and Vogelstein 1997). In the retina, the Rb cell-cycle regulatory gene appears to play the gatekeeper role (Newsham and others 1998); in sunlight-induced nonmelanoma squamous carcinoma of the skin, p53 plays this role (Brash and others 1991); in breast cancer it might be the Brca1 and 2 genes (Couch and Weber 1998); in basal cell cancers it might be the signal-transduction pathway involving genes called "*hedgehog*" and "*smoothened*" (Xie and others 1998; Epstein Jr 1996; Johnson 1996). In the progression of stomach cancer, p53 mutations occur earlier than in the small intestine, and tumors with mutations in EGFR-1 are more aggressive than those with p21 (*WAF1*) mutations. Amplification of c-erbB-2 and of some specific chromosomal regions and loss of heterozygosity in a region containing thymine glycosylase have been reported in stomach cancers.

When a gatekeeper gene can be clearly identified, it should contain mutations or rearrangements that are characteristic of the initiating damage, in that these changes represent some of the earliest genetic events in carcinogenesis. Tissue-specific genes might be similarly involved in the initiation of cancers of lung, stomach, and other tissues by exposure to radon and radon progeny alpha particles. Some of the initial genetic changes resulting from alpha-particle irradiation, such as deletions and rearrangements, are distinctive and might leave characteristic genetic changes, or "fingerprints," on gatekeeper genes and on others activated early by exposures, thus aiding in their identification. But deletions and rearrangements are also common events during tumor progression because of their inherent genomic instability, so alpha-particle fingerprints might be obscured in advanced tumors.

TUMOR GROWTH AND NUTRITION—THE CATERERS

Tumors rapidly outgrow the capacity of diffusion from pre-existing blood supplies to provide the oxygen, nutrients, and growth factors required to sustain their growth and expansion. Anoxic regions of tumors that develop far from blood vessels have been shown to contain elevated amounts of p53 indicative of their abnormal state (Graeber and others 1994). Consequently, a critical factor in tumor growth is the capacity to stimulate new blood vessel growth—angiogenesis. Angiogenesis is achieved by a combination of mechanisms. Tumors secrete stimulators of new blood vessel formation (vascular endothelial growth factors) and reduce the presence of inhibitors (Boehm and others 1997; Folkman 1996). Because the growth of new blood vessels involves the proliferation of essentially diploid, normal endothelial cells, these do not exhibit the genomic plasticity of tumor cells and are subject to normal cell regulation. Proliferative endothelial cells exhibit characteristics and gene-expression profiles different from those of mature established blood vessels and so might even constitute a unique target for cancer therapy (Boehm and others 1997).

GENETIC INSTABILITY IN IRRADIATED CELL POPULATIONS— THE DIVERSIFIERS

Damage caused by high- and low-LET radiation exposure appears to create a genetically unstable state in which further chromosomal and genetic changes can be observed many generations after the exposure. That was first observed for alpha particles by Kadhim and others (1994; 1992) who detected chromatid and chromosomal type aberrations in clonal descendants and nonclonal cultures of both mouse and human hematopoietic stem cells. The instability is not confined to high-LET radiation, and it can even be induced by ionization produced outside the nucleus. Abnormal karyotypes were observed several passages after irradiation; this indicated that heritable changes were transmitted to progeny cells and resulted in new chromosomal rearrangements during later cell cycles. There is evidence that those changes can involve a wide variety of genetic events, including rearrangements, gene amplification, and mutation. DNA sequence rearrangements can lead to mutations, the production of new fusion genes, or changes in gene regulation by position effects that are known to be involved in chromosomal activation of oncogenes in several human and rodent malignancies (Rabbitts 1994). The mechanism of instability might involve rearrangements that result in inappropriate gene expression that then triggers later genetic events. Alternatively, it could involve persistent changes in gene expression through p53 and other gene products that act as altered transcriptional regulators.

The high frequency of chromosomal abnormalities and mutations in human cancers indicates that a "mutator" phenotype is often involved in multistep carcinogenesis (Loeb 1994; 1991). The spontaneous-mutation rate in normal diploid cells is insufficient to account for the high frequency of mutations in cancer cells. Rather, the genomes of cancer cells are unstable, and this results in a cascade of mutations that cumulatively enable cancer cells to bypass the host regulatory processes (Loeb 1994). The development of genetic instability, especially the capacity for gene amplification, is acquired in stages through preneoplastic to fully neoplastic cells, and this capacity appears to depend on the progressive loss of p53 function (Tlsty 1996; Tlsty and others 1995).

DNA damage of various kinds is particularly effective in inducing genomic instability, whether produced by α-particles or x rays or endogenously. For example, an anoxia-inducible endonuclease activity has been reported that cleaves DNA without specificity for sequence (Stoler and others 1992). That activity could account for the induction of gene amplification in anoxic cells and could be associated with break-related genomic instability. Repeat sequences, such as interstitial telomere-like repeats might also be hot spots for recombination, breakage, and chromosome fusion (Alvarez and others 1993; Ashley and Ward 1993; Bouffler and others 1993; Day and others 1993; Hastie and others 1990). Even chromosomal rearrangements that appear stable, such as balanced translocations, are not as secure as normal chromosomes and show declines in frequency with time after radiation exposure in vivo (Tucker and others 1997). A frequent result of chromosomal instability in tumor progression is the loss of a chromosome and the reduplication of the homologue; the chromosome number is maintained with the loss of heterozygosity (LOH). That can result in the loss of a normal gene and the duplication of mutant genes. Recent analysis of a large number of tumors indicates that LOH can involve an exceedingly large variety of genome-wide alterations even for a single tumor type (Kerangueven and others 1997).

The processes of mutation, insertion, deletion, rearrangement, loss of heterozygosity, reduced apoptosis, radiation-induced genomic instability, and the continued replication and proliferation of stem cells lead to a number of critical changes in genes along the paths that result in malignancies. Each tissue might require changes in specific genes, possibly in a particular sequential progression, for complete malignancy to emerge. The need for an ordered set of changes leads to the concept of fingerprints: characteristic mutations in tissue-specific rate-limiting genes that need to be altered early to allow tumor progression (Dogliotti 1996).

MUTATIONS IN α-PARTICLE-INDUCED TUMORS—
THE FINGERPRINTS

It would be expected on the basis of in vitro work, that radon alpha-particle-induced cancers of the lung and other tissues would contain characteristic mutations, fingerprints, in critical gatekeeper genes that initiate carcinogenesis (Dogliotti 1996). Genetic changes that occur during tumor progression are likely to involve many genes but would lack characteristic fingerprints. The strongest example of a carcinogenic fingerprint is the detection of C to T mutations at the 3′ C in dipyrimidine sites in nonmelanoma skin cancers, representing mispairing at sites of sunlight-induced photoproducts in DNA (Brash and others 1991). Carcinogenesis, however, is a highly selective process in which many genetic changes are pruned by selective constraints before the fully malignant cell types with genetic variability, unregulated growth, and invasive properties emerge. Many large α-particle-induced deletions might therefore be inconsistent with emergence of these properties and be lost from such a population. Deletions with a range of sizes up to complete gene loss, however, have been observed in the Rb and p53 genes in murine tumors induced by ionizing radiation (Zhang and Woloschak 1997). The mutations observed in α-particle-induced tumors will therefore be a subset of the full spectrum of genetic changes that are produced initially. But with the prevalence of genomic instability in tumors, specific deletions and rearrangements might easily become obscured; this would make a search for α-particle-induced fingerprints difficult.

A potential fingerprint of α-particle damage at the whole-chromosome level has been suggested. Because of the physical distribution of α-particle tracks, there might be a much lower ratio of interchromosomal exchange aberrations to intrachromosomal exchanges compared to the ratio of these exchanges induced by either low-LET radiation or chemically induced damage (Brenner and Sachs 1994). Again, after the scrambling of the genotype associated with tumor progression, this ratio would be extremely difficult to assess in advanced tumors.

There have been only a few analyses of tumors known to be induced by radon or other α-particle exposure. One set of results is from miners who experienced high radiation doses and dose rates—doses that might not correspond to the exposures expected from domestic situations. The evidence of signature mutations is not strong. A report described point mutations in codon-249 and 250 of the p53 gene, but these could have been spontaneous events or induced by the molds or cigarette-smoking associated with miners' working conditions (Vahakangas and others 1992). The presence of signature mutations in the p53 gene therefore remains to be established. Alternatively, if p53 is not the critical, rate-limiting gatekeeper gene for lung carcinogenesis, signature mutations might yet be identified when the appropriate genes are known and investigated.

EPIDEMIOLOGIC, BIOPHYSICAL, AND CELL-BASED MODELS OF RADON-INDUCED CARCINOGENESIS

To obtain estimates of risk posed by exposure to radon in air or drinking water, it would be ideal to trace the complete process from α-particle exposure to cancer, on a quantitative, biologic, and molecular basis and to incorporate such difficult issues as individual and subpopulation variations in susceptibilities (see BEIR VI, National Research Council 1999). Unfortunately that is not yet feasible. Instead, the problem of risk estimation has been approached from a variety of avenues. One is through strict epidemiologic relationships between numbers of cancers and exposure and the use of the linear no-threshold dose-response curves used commonly in radiation risk estimates. Another approach introduces biophysical models of radiation action based on radiation tracks, total doses and dose rates, damaged sites in DNA, and breaks and their rejoining and, from these considerations, reaches interpretations of risk versus dose. Most often, this approach has been used to explain such phenomena as the inverse dose-rate effects relevant at doses higher than those that would arise from domestic exposures (Brenner 1994; Elkind 1994). The approach still does not pay specific attention to the particular genetic changes involved in cancers, and a more detailed attempt to interpret carcinogenesis on a quantitative basis has incorporated changes in cell cycles, proliferation kinetics, cell-killing (Luebeck and others 1994), and other biological processes alluded to in this chapter (Luebeck and others 1994). Because of the large numbers of variables involved, these approaches are still too difficult, computationally, to incorporate into a completely predictive model.

Instead, the biologic approach gives a mechanistic underpinning to the epidemiology and biophysical interpretations of risk. Together, they lead to a more comprehensive understanding of cancer risks posed by low ambient radiation exposures and provide a rational basis for quantitative exposure risk assessment and mitigation by multimedia approaches to risk reduction.

7

Defining Key Variabilities and Uncertainties

Estimating potential human exposures to and health effects of radon in drinking water involves the use of large amounts of data and models for projecting relationships outside the range of the observed data. Because the data and models must be used to characterize population behaviors, engineered system performance, contaminant transport, human contact and dose-response relationships among different populations in different geographic areas, large uncertainties and variabilities are associated with the resulting risk characterization. In this chapter, the committee evaluates the importance of and methods for addressing uncertainty and variability that arise in the process of assessing multiple-route exposures to and health risks associated with radon. The data, scenarios, and models used to represent human exposures to radon in drinking water include at least five important relationships:

- The magnitude of the source-medium concentration, that is, the concentration of radon in the water supply or in ambient air.
- The contaminant concentration ratio, which defines how much a source-medium concentration changes as a result of transfers, transformation, partitioning, dilution, and so on before human contact.
- The extent of human contact, which describes (often on a body-weight basis) the frequency (in days per year) and magnitude (in liters per day) of human contact with a potentially contaminated exposure medium.
- The duration of potential contact of the population of interest as related to the fraction of lifetime during which an individual is potentially exposed.
- The averaging time for the type of health effects under consideration; for example, the appropriate averaging time could be a cumulative duration of expo-

sure (as is typical for cancer and other chronic diseases) or it could be a relatively short exposure period (as is the case for acute effects).

On the basis of those five relationships, figure 7.1 illustrates the steps of the risk-assessment process for multimedia human exposure to radon. The emphasis in the figure is on the outcome calculated at each step and the types of data needed to calculate the outcomes.

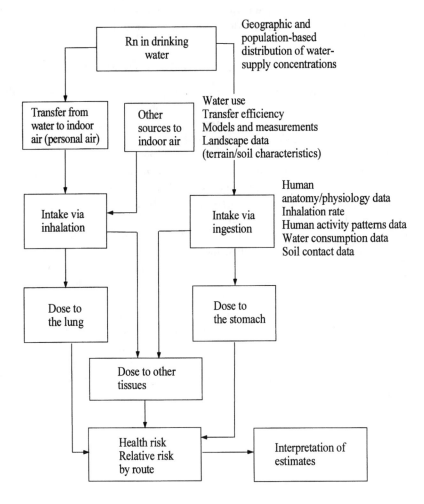

FIGURE 7.1   Steps of risk-assessment process for multimedia human exposure to radon, with emphasis on outcome calculated at each step and types of data needed to calculate outcomes.

This chapter begins with an overview discussion about factors that determine the reliability of a risk assessment and a discussion of methods for characterizing and evaluating the uncertainties in a risk assessment. Next is a summary review and evaluation of the uncertainty analysis for drinking-water radon that was carried out by the Environmental Protection Agency. That is followed by the committee's consideration of the steps of the risk-assessment process described in earlier chapters and of how uncertainty and variability apply to the assessment and the extent to which they can be quantified. Particular attention is given to the importance of uncertainty across the entire process of characterizing the unmitigated risk associated with radon in drinking water and the risk reduction achieved by various technologies used to reduce radon levels in water supplies.

RELIABILITY OF A HEALTH-RISK ASSESSMENT

To identify factors that affect the reliability of radon risk assessment, the committee reviewed the scientific literature, recommendations from other National Research Council studies, and findings reported by such organizations as the International Atomic Energy Agency (IAEA), the National Council on Radiation Protection and Measurements (NCRP), and the Presidential/Congressional Commission on Risk Assessment and Risk Management. According to IAEA (1989), five factors determine the precision and accuracy, that is, the reliability, of a risk characterization: specification of the problem (scenario development), formulation of the conceptual model (the influence diagram), formulation of the computational model, measurement or estimation of parameter values, and calculation and documentation of results, including uncertainties. In such a framework, there are many sources of uncertainty and variability—including lack of data, natural-process variation, incomplete or inaccurate data, model error, and ignorance of the relevant data or model structure.

The magnitude of human exposure to toxic agents, such as radon, often must be estimated with models that range in complexity from simple heuristic extrapolations from measured trends to large-scale simulations carried out on large computers. Regardless of its complexity, any model can be thought of as a tool that produces an output, Y, such as exposure or risk, that is a function of several variables, X_i, and time, t:

$$Y = f(X_1, X_2, X_3, \dots X_k, t). \qquad (7\text{-}1)$$

The variables, X_i, represent the various inputs to the model, such as radon concentration in water, and the transfer factors between water and air. Uncertainty analysis involves the determination of the variation or range in the output-function values—that is, risk values—on the basis of the collective variation of the model inputs. In contrast, a sensitivity analysis involves the determination of the changes in model response as a result of changes in individual parameters. An approach to express the combined impact of uncertainty and variability more fully is to perform a two-dimensional Monte Carlo simulation consisting of an inner set of calculations embedded within an outer set. That was first described by Bogen and Spear (1987).

One of the issues that must be confronted in uncertainty analysis is how to distinguish between the relative contributions of variability (heterogeneity) and true uncertainty (measurement precision) to the characterization of predicted outcome. *Variability* refers to quantities that are distributed empirically—such factors as rainfall, soil characteristics, weather patterns, and human characteristics that come about through processes that we expect to be stochastic because they reflect actual variations in nature. These quantities are inherently random or variable and cannot be represented by a single value, so we can determine only their characteristics (mean, variance, skewness, and so on) with precision. In contrast, true *uncertainty,* or model-specification error (such as statistical estimation error), refers to an input that, in theory, has a single value, which cannot be known with precision because of measurement or estimation error.

Uncertainty in model predictions arises from a number of sources, including specification of the problem, formulation of the conceptual model, estimation of input values, and calculation, interpretation and documentation of the results. Of the factors that determine precision and accuracy, only uncertainties due to estimation of input values can be quantified in a straightforward manner on the basis of variance propagation techniques. Uncertainties that arise from mis-specification of the problem and model-formulation errors can be assessed using less straightforward processes, such as decision trees and event trees based on expert opinions. In some cases, using such methods as meta-analysis, model-specification errors can be handled with simple variance-propagation methods.

ENVIRONMENTAL PROTECTION AGENCY PROCESS FOR ASSESSING AND EVALUATING UNCERTAINTIES IN RADON RISK

In support of its proposed rule for radionuclides in drinking water, EPA has developed estimates of the cancer risk associated with radon in drinking water. The risk arises from multiple exposure pathways, including the direct ingestion of water that contains radon, the inhalation of indoor air that contains radon some of which has volatilized from water used in the home, and the inhalation of radon progeny that are introduced into indoor air as a result of radon decay. Because exposure and dosimetry are different for each pathway, EPA has estimated the risks associated with radon in drinking water by calculating the risk for each pathway separately and then combining risk to obtain the total risk related to all pathways. In an earlier risk assessment (EPA 1995), EPA estimated the total (all-pathways) average lifetime risk to the US population posed by radon in drinking water as 6.6×10^{-7} per picocurie per liter of radon in water.

After the risk estimates were performed, EPA obtained new data on radiation dosimetry that required revision of the estimates for radon in drinking water. The new data included a National Research Council re-evaluation of the relative dosimetry of radon decay products in mines and homes (National Research Council 1991b). On the basis of the new data, the total (all-pathways) lifetime risk estimate for radon was changed to 7×10^{-7} per picocurie per liter. Although the total risk by pathway did not change substantially, the allocation of risk contributed by each pathway did. However, perhaps a more important component of the revised risk numbers was the inclusion of a detailed uncertainty analysis (EPA 1995). Appendix F provides the committee's summary and evaluation of the EPA uncertainty analysis. Our focus is on how the explicit analysis of uncertainty and variability can influence the process for setting standards and for setting priorities for intervention and future research.

In reviewing the EPA uncertainty and variability analysis, the committee found that the EPA approach demonstrated innovative methods and consistency with emerging policies, and provided an adequate characterization of the uncertainty in cancer risk factors. However, our analysis of and proposed revision in the risk models will result in changes in both the magnitude and the uncertainty ranges of some of the parameters in the EPA model. In particular, the magnitude of the risk associated with radon ingestion has been lowered in the committee's analysis, but the resulting uncertainty range is contained within the uncertainty range used by EPA (1995).

Although the EPA analysis was an important initial effort at uncertainty assessment, results of that analysis can be misleading. In particular, because the variability in the risk-per-dose factors cannot be specified, the variability in risk derived from this analysis includes only variability in exposure and not the actual variability in cancer risk among the population. Moreover, in reviewing the EPA models and uncertainty analysis, the committee observed that implicit in the development of this model is the assumption that the risk factor is independent of variability in the unit dose factor. That assumption requires that the radon-gas dosimetry be independent of the breathing rate—an assumption that is not consistent with the key issues of inhalation dosimetry described in chapter 5 of this report.

The committee had a particular interest in the radon-ingestion risk model because the ratio of ingestion risk to inhalation risk is an important component of the multimedia approach to radon risk management. The committee observed that the EPA risk assessment used an appropriate approach to obtain the uncertainty factor for the population cancer risk associated with ingestion of radon in water. EPA assigned a geometric standard deviation of 2.4 to the risk factor for ingestion-cancer risk. That implies that there is a 68% likelihood that the actual risk factor is within a range of roughly 2.4 times lower to 2.4 times higher than the estimated risk and a 95% likelihood that the actual risk factor is within a range of roughly 6 times lower to 6 times higher than the estimated risk factor. That uncertainty range reflects parameter uncertainty associated with the risk model used by EPA. However, the model is constrained by the assumption that radon is instantaneously and uniformly distributed in the stomach after ingestion. The uncertainty range in the EPA results does not appear to reflect the bias and uncertainty associated with this assumption. Nevertheless, the 95% confidence interval of the uncertainty range developed by the committee for this report (in chapter 4) is essentially contained within the 99% confidence interval suggested by the EPA results. This reveals that the EPA did not underestimate their confidence interval. However, this committee's uncertainty range is at the lower bound of the EPA range, suggesting the likely upward bias of the EPA risk estimate.

The EPA risk assessment made no effort to assess the contribution of soil relative to that of water to indoor radon levels. Therefore, the study affords little input to the analysis of how any standard or policies can affect the risk associated with all radon exposures. Uncertainty of the ingestion:inhalation risk ratio is an important factor that was not addressed in the EPA analysis.

ISSUES IN UNCERTAINTY ANALYSIS FOR RADON

For an agent like radon, which is ubiquitous, total exposure might reflect concurrent contacts with multiple media instead of continuous or multiple contacts with a single medium. Multimedia pollutants give rise to the need to address many types of "multiples" in the quantification or measurement of exposure and dose, such as the multiple media themselves (air, water, soil): multiple exposure pathways (or scenarios), multiple routes (inhalation, ingestion, and dermal), and multiple exposure target tissues for dose and effect.

There are many sources of uncertainty and variability in the process of exposure and human-health assessment. The variability and many of the uncertainties cannot be reduced. One common approach to addressing uncertainty in exposure and risk assessments is contrary to the accepted principles of decision-making in the presence of uncertainty. This is the practice of compounding upper-bound estimates as a means of basing decisions on a highly conservative estimate of exposure. Such compounding of upper-bound estimates leaves a decision-maker with no flexibility to address margins of error, to consider reducible versus irreducible uncertainty, to separate individual variability from true scientific uncertainty or to consider benefits, costs, and comparable risks in the decision-making process. Because the compounding of conservative estimates does not serve the exposure-assessment process well, there is a growing effort to include uncertainty analyses in the risk assessment process. EPA has taken the latter uncertainty-analysis approach in its risk assessment for radon, and the committee believes that it is important to continue this precedent.

For human populations, total-exposure assessments that include time and activity patterns and microenvironmental data reveal that an exposure assessment is most valuable when it provides a comprehensive view of exposure pathways and identifies major sources of uncertainty. In any issue involving uncertainty, it is important to consider a variety of plausible hypotheses about the world, to consider a variety of possible strategies for meeting goals, to favor actions that are robust to uncertainties, to favor actions that are informative, to probe and experiment, to monitor results, to update assessments, and to modify policy accordingly and favor actions that are reversible (Ludwig and others 1993).

To make an exposure assessment consistent with such an approach, both sensitivity and uncertainty analyses should be incorporated directly into an iterative process in which premises lead to measurements, measurements lead to models, models lead to better premises, better premises lead to additional but better-informed measurements, and so on. In 1996, the EPA Risk Assessment Forum held a workshop on Monte Carlo analysis. Among the many useful discussions at the meeting was a call for a "tiered" approach to probabilistic analysis, which is iterative and progressively more complex. The need for formal uncertainty analysis and a tiered approach will require the development by the exposure-assessment community of new methods and will put greater demands on the number and types of exposure measurements that must be made. At least three tiers are needed, as follows:

- First, the variances of all input values should be clearly stated, and their effect on the final estimates of risk assessed. At a minimum, that can be done by listing the estimation error or the experimental variance associated with the parameters when these values or their estimation equations are defined. It would help to define and reduce uncertainties if a clear summary and justification of the assumptions used for each aspect of a model were provided. In addition, it should be stated whether the assumptions are likely to result in representative values or conservative (upper-bound) estimates.
- Second, a sensitivity analysis should be used to assess how model predictions are affected by model reliability and data precision. The goal of a sensitivity analysis is to rank input parameters on the basis of their contributions to variance in the output.
- Third, variance-propagation methods (including but not limited to Monte Carlo methods) should be used to map how the overall precision of risk estimates is tied to the variability and uncertainty associated with the models, inputs, and scenarios.

THE COMMITTEE'S EVALUATION OF UNCERTAINTIES IN RISK ASSESSMENT OF RADON IN DRINKING WATER

Uncertainties in Molecular Biology of Cancer Induction by Radiation

As discussed in chapter 6, the exposure of human cells to the high-LET radiation from the decay of radon and its progeny initiates a series of events that can lead to lung and other cancers. This series of events is now thought to be well outlined, but the quantitative link between radon concentration in tissues and cancer risk cannot yet be derived from a quantitative analysis of these processes. Before estimates of cancer risk posed by radon in air and drinking water can be based on quantitative models of biologic and molecular processes, these models must incorporate such difficult issues as individual and subpopulation variations in susceptibility. However, because of the large number of variables involved in such models, and the lack of detailed understanding of each step in the process, the models are still far too difficult computationally to be used for radiation risk assessment. For the near future, risk estimates must be based on current quantitative epidemiologic relationships between numbers of cancers and exposure in selected high-exposure populations. Neither this committee nor the BEIR VI committee has made risk estimates based directly on the emerging biophysical and cellular models.

However, the study of molecular and cellular mechanisms of radiation-induced cancer brings to the risk assessment process important insights about the nature and magnitude of the uncertainties associated with the dose-response models discussed in this report. In particular, the introduction of biophysical cellular models to the risk assessment process reveals both the limited reliability and potential bias of the existing risk assessment models. Biophysical models relate the amount and persistence of biological damage to factors such as radiation tracks, total doses and dose rates, damaged sites in DNA, and DNA breaks and their rejoining. These models can be used to explore inverse dose-rate effects and some of the age-variation in effects. Cellular models focus on changes in cell cycles, proliferation kinetics, cell killing, cell regulation, and other processes that alter the path from radiation deposition to cancer incidence. Although still in the early phase of development, these models may eventually be used to explore variations in susceptibility associated with age, gender, and other genetic characteristics. Nevertheless, these emerging models and the mechanisms of action being studied by radiation biophysicists have provided this committee and others guidance for estimating the uncertainties associated with dose-response functions. Perhaps the most important insight is the recognition of the uncertainty regarding the relevance of the population used to develop a dose-response model. Radon risk derived from a particular population, such as survivors of the atomic bombings of Hiroshima and Nagasaki or miners, cannot necessarily be used directly to estimate risk for a different population such as the US population exposed to radon. This inability to transfer the risk estimates occurs because radiation-induced cancer risk is a function of the underlying spontaneous-cancer incidence (see for example National Research Council 1990a). Average risk estimates are obtained from epidemiology studies that can detect radiation-induced effects in large groups. The problem of extrapolating from one population to another is often dealt with by assigning an appropriate uncertainty interval to the risk estimates. To assign appropriate uncertainties, however, there is a need for more detailed data for which the distribution of risks among individuals can be determined.

Uncertainty in Ambient Radon Levels

With regard to uncertainty and variability about radon levels in ambient air and groundwater, there are several key questions, including:

- What is the variation of radon levels in soil, groundwater, and drinking water in the United States? Not only must consideration be given to variation but also to how reliably it can be characterized on the basis of the number and geographic extent of the available measurements.
- What is the distribution in the US population of radon levels in water supplies and in the soil adjacent to residences? The issue here is to develop population-weighted distributions of radon levels in soil, indoor air, and in water supplies. Of particular importance is the joint distribution of radon levels in soil and water at the high end of their respective distributions. Because such joint probability distributions are not well characterized, constructing them involves judgment, assumptions, and approximations that will introduce uncertainty.

National data on indoor radon, radon in water, and geologic radon potential indicate systematic differences in the distribution of radon across the United States. From geologic-radon potential maps and from statistical modeling of indoor radon exposures, it is clear that the northern United States, the Appalachian and Rocky Mountain states, and states in the glaciated portions of the Great Plains tend to have higher than average indoor radon (see chapter 2).

Available data on radon in water from public water supplies indicate that higher concentrations of radon in water occur in the New England, Appalachian, and Rocky Mountain states and in small areas of the Southwest and Great Plains. Available data also indicate that small water supplies have higher average radon concentrations than large ones. The reasonable agreement of water concentration variation among the various studies suggests that the Longtin (1990) data used by EPA (1995) are adequate for representing variations in water-supply radon concentrations.

The ambient concentration of radon outdoors varies with distance and height from its principal source in the ground (rocks and soil) and from other sources that can locally or regionally affect it, such as bodies of water, mine or mill tailings, vegetation, and fossil-fuel combustion. However, diurnal changes due to air stability and meteorologic events account for most of the variability. As reported in chapter 2 of this report, the committee does not believe that the available data are sufficiently representative to provide a population-weighted annual average ambient radon concentration. From the available data, the committee has obtained an unweighted average of 15 Bq m^{-3} with a standard error of 0.3 Bq m^{-3}. The committee recommends this value as the best available national average ambient concentration. In reviewing all the other ambient-radon concentration data that are available for other specific sites, the committee concluded that the average ambient radon concentration would most likely be 14-16 Bq m^{-3}. Thus, it is the committee's recommendation to treat the value of the average ambient radon concentration as being represented as a uniform distribution of range 14-16 Bq m^{-3} with a most probable value of 15 Bq m^{-3}.

Variability and Uncertainty in Transfer Factors

The committee considered and re-evaluated the variability in the transfer of radon gas from water to indoor air. Assessing the increment of airborne radon in a home that arises from the use of water that contains dissolved radon is a problem that involves both uncertainty and variability. It involves the solubility of radon in water, the amount of water used in the home, the volume of the home, and the home ventilation rate. The amount of radon from the water is not constant throughout a home, but is higher in areas of active water use, such as bathrooms and kitchens. Table 7.1 summarizes the recommended values of the transfer factor and the parameters used to construct it.

The resulting geometric mean value is 5.5×10^{-5} or 3.9×10^{-5} with a geometric standard deviation (GSD) of 3.5. These values can be compared with those of Nazaroff and others (1987) who reported a geometric mean of 6.5×10^{-5} and a GSD of 2.8, and EPA (1995), which reported a geometric mean of 6.5×10^{-5} and a GSD of 2.9. There was reasonable agreement between the geometric mean of the transfer coefficient estimated by the model and the estimated value calculated from the measured data. The average of the measurements was 8.7×10^{-5} with a standard error of 1.0×10^{-5}. With the modeled geometric mean ventilation of 1.07 air changes per hour, the calculated transfer coefficient is the same value as the measurements. However, if we use the estimate of the geometric mean of the ventilation rate of 0.77, the resulting estimate of the transfer coefficient is 1.2×10^{-4}. The committee feels that there are problems with both the measurements of the transfer coefficient and the measurements that are the input values into the model. The committee recommends that EPA continue to use 1.0×10^{-4} as the

TABLE 7.1 Parameters of the Lognormal Distributions for the Parameters in the Transfer-Factor Calculation

Parameter	Committee's Values	
	Geometric Mean	Geometric Standard Deviation
House volume per occupant, m^3 person^{-1}	115	2.0
Ventilation rate	0.77 or 1.07	2.3
Transfer efficiency	0.52	1.3
Water use per capita, m^3 person^{-1} hr^{-1}	$9.4 \ 10^{-3}$	1.8
Transfer coefficient	5.5×10^{-5} or 3.9×10^{-5}	3.5

best central estimate of the transfer coefficient that can now be obtained. Further, because of the uncertainty in the value of the ventilation rate and its distributional characteristics, the committee recommends that the transfer coefficient be assumed to be in the range $0.9\text{-}1.2 \times 10^{-4}$. The committee is not assigning a specific uncertainty to the central estimate, but rather suggesting that it has the highest likelihood of lying within this range.

Those are not particularly large changes, suggesting that because the parameters have remained stable even as the amount of data relating to the transfer factor has increased, the uncertainty might now be lower than that suggested by the EPA analysis (EPA 1995). Further studies on transfer factors will not reduce the uncertainty substantially. The committee did not find a compelling need to go to a three-compartment model; it is not particularly more effective in characterizing either the uncertainty or the variability of the transfer-factor calculation.

An important issue is the integration of the dosimetry model with the model of radon-progeny buildup in the bathroom and in the rest of house volume after water uses. More studies on the buildup and distribution of radon progeny will have much more importance with regard to the overall uncertainty in the link between the concentration of radon in water and inhalation dose. These issues have been discussed in more detail in chapter 5.

One important issue regarding the transfer factor is the question of whether there is a correlation of the distribution of variability and uncertainty in the transfer factor with the distribution of ambient radon levels. For example, there is a need to consider further whether there is a joint occurrence of high radon-in-water levels with geographical regions with high temperature so that both increased tapwater intake and higher radon-in-water concentrations might correspond. Similarly, there is the question of whether high radon-in-water levels occur in regions with low annual temperatures and more tightly sealed homes so that the high radon levels in water would yield to the higher water-to-indoor-air transfer factors.

Inhalation Risk per Unit of Radon-in-Water Concentration for Inhalation

The committee did not conduct its own detailed uncertainty analysis for the risk model used for radon inhalation. Instead it reviewed the uncertainty analyses that have been carried out previously by EPA (1995) and by the BEIR VI committee (National Research Council 1999) to estimate the uncertainties associated with inhalation exposures.

As has been noted by the BEIR VI committee, it is not feasible to conduct a complete quantitative analysis of all potential sources of uncertainty and variability in the estimate of the lung-cancer risk associated with the inhalation of radon and its progeny. A key limitation of such an analysis is the difficulty in enumerating all factors that could influence the lung-cancer risk associated with indoor exposures to radon. An additional limitation is that existing information does not support a fully quantitative characterization of the uncertainty and variability in some of these factors.

The BEIR VI committee focused its quantitative uncertainty analysis on the population attributable risk (AR) associated with radon. Because the AR is a measure of population risk rather than of individual risk, the variability among individuals was not quantified in the BEIR VI analysis. The uncertainty analysis was applied in BEIR VI to the BEIR VI committee's two preferred models—the exposure-age-concentration model and the exposure-age-duration model. BEIR VI uncertainty factors reflecting only uncertainty in the parameters of the BEIR VI risk models provide the geometric range of uncertainty associated with the BEIR VI model. For males, the ratio of the high to low values in the 95% confidence interval of AR is 2.7 for the exposure-age-concentration model and 2.3 for the exposure-age-duration model. The ratios are similar for females. On the assumption that those uncertainty ranges can be represented by log normal distributions, the BEIR VI committee derived from these ratios a GSD of approximately 1.3 for the exposure-age-concentration model and 1.2 for the exposure-age-duration model. From those results, we select an uncertainty factor of 1.3 to be applied to the inhalation risk factor for situations when the equilibrium factor used by BEIR VI applies.

There remains inadequate information to measure and characterize inter-individual variability in the inhalation-risk models that are available for this study. As a result, the cancer-risk models for inhalation described in the BEIR VI report are characterized only in terms of uncertainty, not of variability. However, when the AR is used as a measure of population rather than individual or sub-population risk, the inter-individual variability in cancer risk is effectively averaged out in the analysis. A problem arises when population-based risk factors are applied to small populations or individual households (such as a small number of houses with high radon). In such cases, the failure to know the appropriate risk factors for this small population—where interindividual, variability may not average out—constitutes an important uncertainty.

Ingestion Risk per Unit of Water Concentration for Ingestion

One of the important uncertainties in our analysis involved the issue of radon gas behavior in the stomach. During the information-gathering phase of our analysis, the committee heard conflicting information about the potential of inert gases, such as radon, to be transferred from the contents of the stomach through the mucus layer and to the stem cells surrounding the stomach. The extent to which radon is transferred into and through the stomach wall has a large effect on the predicted radiation dose associated with water ingestion. Previous efforts were based on assumptions that either there was no diffusion through the stomach wall or that the entire stomach wall contains radon at the same concentration as the stomach contents (see chapter 4 for more discussion). These bounding assumptions lead to disparate results regarding the estimated risk. To confront this issue of uncertainty better, the committee elected to develop a stomach model that allowed exploration of a range of diffusion conditions in the stomach and a model characterizing the behavior of radon dissolved in blood and body tissues. Once the radon has entered the blood, through either the stomach or the small intestine, it is distributed among the organs of the body according to the blood flow to the organs and the relative solubility of radon in the organs and in blood. Radon dissolved in blood that enters the lung will equilibrate with air in the gas-exchange region and is removed from the body; this model is described in detail in chapter 4 and in Appendixes A and B. The dosimetry model indicates that any radon absorbed in the stomach results in a higher risk per Bq than in the intestines. The need for the new models also arose from the lack of directly applicable experimental observations and from limitations in the extent to which one can interpret results of existing studies. Risk relevant to ingestion of radon in water depends heavily on the extent to which radon penetrates the stomach wall. With the new model, the committee was able to conduct a broader set of sensitivity and uncertainty analyses. The committee notes that limitations in the model structure with regard to the relative locations of the microvasculature structure (and its fractional capture of the diffusing radon) and stem cells are the major sources of uncertainty. The diffusion of radon within the stomach wall was modeled to determine the expected time-integrated concentration of radon at the depth of the cells of risk in the stomach wall. The committee's baseline (or median) estimate is based on a radon diffusion coefficient of 5×10^{-6} cm^2/s. Using this value yielded an integrated radon concentration in the wall that is about 30% of the concentration in the contents of the stomach. Sensitivity and uncertainty analyses with this model helped the committee to bracket the range of risks that could plausibly be associated with ingestion of radon in water. The committee estimated that the diffusion coefficient in the stomach could have a plausible lower bound of 10^{-7} cm^2/s and a plausible upper bound of 10^{-5} cm^2/s (the diffusion coefficient of radon in water). That range of diffusion coefficients results in a median estimate of risk of 2.0×10^{-9} per becquerel per m^3. However, the "no diffusion" and "saturated diffusion" limits in the calculations which were carried out were not intended as realistic limits and thus should not be interpreted as representing the range of uncertainty in the ingestion risk. This range was selected to reflect the current literature, in which some authors believe that diffusion is not a viable mechanism, others avoided the whole issue, and still others endorsed it with the intent of being conservative when setting a radiation-protection quantity. The committee's calculations in the extremes were largely for the purpose of illustrating the significance of this mechanism; that is, they were bounding calculations.

The committee has not carried out a detailed uncertainty analysis for the ingestion-risk model described in chapter 4. However, it has made some subjective judgments regarding uncertainties on the basis of what has been done and what is now known about this problem. From this review, the committee makes the following observations:

- The literature on inert gases in the stomach clearly supports an assumption that movement into the stomach wall occurs, and diffusion is the probable mechanism.
- Physiologic processes and histologic structures prevent gastric acids from digesting the stomach, and it is reasonable to assume that they restrict to some degree the movement of gases into the wall.
- The basic input data for the calculations of stomach-cancer risk are based on risk factors derived from the Japanese atomic-bomb survivors, and a high background of stomach cancer among the Japanese population is well established. Thus, the Japanese data are transported to the US population with a relative-risk projection model that considers the background rate of stomach cancer in the United States. The incidence of stomach cancer in the US population involves a number of cofactors and has been declining in recent years.

It is the judgment of the committee that the risk of cancer posed by an absorbed dose in the stomach is probably not greater than 2.3 times the best estimate of 1.6×10^{-9} per Bq m^{-3} and it is probably greater than this value divided by 5; that is, it is probably between 3.8×10^{-10} to 4.4×10^{-9} per becquerel per m^3. Assuming that these bounding values represent the 80% confidence interval— that is, a 3.3 standard-deviation range of risk, the committee estimates that the uncertainty in this risk factor has a GSD of 2.1, which is lower than the EPA-estimated GSD of 2.4. Thus, the proposed committee model gives an estimate of risk that is about a factor of 3 lower than the EPA median risk estimate and has a lower GSD that reflects uncertainty. Variations in ingestion are incorporated into this estimate of risk, but uncertainties in the nature and magnitude of diffusion processes in the stomach are dominant contributors to overall uncertainty.

Uncertainty and Variability with Regard to Mitigation

A key issue of uncertainty is quantification of the reduction in the level of radiation dose achieved by various mitigation technologies and how this reduction is distributed among the populations at risk. The actual performance of these technologies, compared with what it is assumed to be, is probably an important uncertainty. Variations in performance and reliability might be large and difficult to quantify.

COMMUNICATION OF UNCERTAIN RISK INFORMATION

The decision to expend societal resources to identify, estimate, and manage risk implies a valuation of the risk being controlled. Because of the inherent uncertainty in risk characterization and risk management, it is important to consider how individuals and societies value uncertain adverse consequences. The committee expects such valuations to be expressed in terms of relative preferences, economic preferences, or ethical constraints. One issue that must be considered when there are large uncertainties in risk estimates is how to communicate this information to the affected public. The committee has found that it is often difficult to find the appropriate language for communicating and discussing uncertainties among ourselves. Thus, it is concerned that this difficulty will be amplified significantly when there is a need to communicate information about uncertainties to the less technically oriented community groups that must make decisions based on relative-risk estimates.

Another National Research Council report, *Improving Risk Communication* (National Research Council 1989), has reviewed a number of issues related to risk communication. That report provides some discussion regarding the problems of uncertainty and variability in risk communications. It is suggested that it is dangerous to quantitatively describe the uncertainties in risk messages. It is generally not possible to describe the complexity of the uncertainty analysis. However, it should be made available in ancillary documents. A key goal of their recommendations is to help audiences distinguish areas of scientific agreement amid what may appear as vast areas of policy disagreement. Careful delineation of existing scientific uncertainty is that it gives audiences a sense of the degree of scientific consensus and allows them to distinguish minor from major uncertainties. Thus, the description of uncertainties is both an essential and difficult part of the communication of risk to the public.

Communicating successfully with the public and with water utilities concerning the uncertainty of the risks of radon in air and in water, as well as the uncertainties regarding the likely benefits of risk-reduction strategies will require involving those parties in the process much earlier than was done previously. Risk managers have previously viewed risk communication as a one-way, temporal educational process in which experts pronounced and the audience listened and learned. An unsuccessful risk communication effort was blamed on audience failure to assimilate the information and act appropriately.

That unilateral approach is being replaced by a multilateral one as agencies with a communication mission work to involve the audience in planning and execution. Serving as an important catalyst for this change was the previously mentioned National Research Council (1989) report which declared "risk communication should be a two-way street," between experts and various groups, that should exhibit a spirit of open exchange in a common undertaking rather than a series of "canned" briefings restricted to technical "nonemotional" issues and an "early and sustained interchange that includes the media and other message intermediaries."

Though involvement of the public is understood to enhance communication with those directly involved, relatively little understanding exists on how to best express risk information and its related uncertainties to broad, non-technical audiences who may have little contact time with the subject. For the various reasons discussed, communication about the uncertainty of the risks calculated here will remain as one of the most challenging of endeavors.

DISCUSSION AND RECOMMENDATIONS

The committee identified the issues of uncertainty and variability as likely to have important scientific and policy implications for the health effects attributable to radon in drinking water. One overarching issue is how uncertainty and variability can affect the reliability of estimated health effects of a given standard and the health benefits of alternative standards and control strategies.

The approach used in the EPA uncertainty analysis, which is summarized in Appendix F, was fully consistent with emerging EPA guidelines and protocols for uncertainty analysis. Moreover, the EPA document, which transmitted these results, has defined the state of the art for uncertainty analysis within EPA. The explicit separation of uncertainty and variability and the resulting two-dimensional Monte Carlo analysis express uncertainty and variability separately on the same graph. Those methods are innovative and useful for understanding the distribution of risk among populations and the impact of various mitigation strategies.

In reviewing the EPA effort, the committee observed that the EPA risk assessment used an appropriate approach to obtain the uncertainty factor for the population cancer risk associated with ingestion of radon in water. The uncertainty range used by EPA reflects parameter uncertainty associated with the EPA risk model. However, this model is constrained by the assumption that radon is instantaneously and uniformly distributed in the stomach after ingestion. The uncertainty range in the EPA results does not appear to have been set up to reflect the bias and uncertainty associated with this assumption. Nevertheless, the 95% confidence interval of the uncertainty range developed by the committee for this report is essentially contained within the 99% confidence interval suggested by the EPA results. That suggests that the EPA did not underestimate its confidence interval. However, the committee's uncertainty range is at the lower bound of the EPA range, and this suggests the likely upward bias of the EPA risk estimate.

Because current risk models must rely on epidemiologic relationships, it is difficult to accurately represent individual and subpopulation variations in susceptibility. However, the study of molecular and cellular mechanisms of radiation-induced cancer brings to the risk-assessment process important insight about the nature and magnitude of the uncertainties associated with the dose-response models discussed in this report. In particular, the introduction of biophysical cellular models to the risk-assessment process reveals both the limited reliability and the potential bias of the existing models.

One critical issue in defining the potential risk associated with waterborne radon is the rate of diffusion of radon through the stomach wall toward the stem cells surrounding it. This rate is critical in defining the relative importance of the risks associated with waterborne radon compared with the risks associated with indoor airborne radon.

There remains insufficient information to quantify interindividual variability in the cancer-risk models that are available. As a result, the cancer-risk models for inhalation described in this report are characterized only in terms of uncertainty, not variability. In contrast, radon exposure data—including concentrations in water and in indoor air, transfer factors, and equilibrium factors—have been collected with sufficient resolution to explicitly represent population variability within the United States. However, uncertainties in the parameters—that is, distributional moments—describing this variability are not yet known with precision. The uncertainty in the parameters describing exposure variability can be measured with methods used by EPA (1995) and Rai and Krewski (1998), which use combined uncertainty and variability analysis to characterize the relative importance of the two sources of variance in risk estimate.

8

Mitigation

MITIGATION OF RADON IN INDOOR AIR

Radon Entry into Buildings: A Brief Review

Radon is a ubiquitous constituent of soil gas as its radioactive parent, ^{226}Ra, is widely distributed in the earth's crust. Typical soil-gas radon concentrations are around 30,000-300,000 Bq m^{-3}, and values ranging from about 5,000 Bq m^{-3} to about 5,000,000 Bq m^{-3} have been reported. The principal mechanisms of radon transport in porous media (e.g., soil) are advection and diffusion; both are sources of radon entry into buildings, and they are described briefly in this section. More complete discussions of radon transport in soils and entry into buildings can be found in the literature (Sextro 1994; Nazaroff 1992; Nazaroff and others 1988).

Advection

Bulk flow of soil gas that contains radon is the main mechanism of radon entry into buildings. This flow occurs in response to pressure differences between the air in buildings and the air in the adjacent soil. These differences are established by the natural interaction between the building and the surrounding environment and in some cases by the operation of mechanical systems within the building.

The temperature difference between the air in a building and the air outside creates a pressure gradient across the building shell that varies with height along the shell. When the indoor air temperature is higher than the outdoor, the indoor air pressure in the lower parts of the building (e.g., in the basement or the region of the ground-contact floor) is slightly lower than the air pressure in the adjacent soil; conversely, at the top of the building, the pressure gradient across the building shell is reversed, so air flows out of the building. This thermal-stack effect is one of two principal mechanisms responsible for the natural ventilation of buildings; it is sometimes referred to as infiltration.

Wind also creates pressure differences between the inside and the outside of a building. The pressure fields can be complex and depend on the size and shape of the building and the wind direction. The pressure fields also extend into the surrounding soil, increasing the pressure of the soil air on the upwind side and decreasing it on the downwind side of the building (Riley and others 1996). The net effect is usually an airflow out of the top of the building caused by the Bernoulli effect of the wind over the roof or by the reduction in air pressure on the leeward side of the building. In response to the slightly lower air pressure in the building, "makeup" air flows in through openings in the building shell, some of which might provide direct contact with soil air.

The effects of the thermal stack and the wind can independently result in indoor-outdoor pressure differences of about the same size at the lower portions of the building shell. For example, an indoor-outdoor temperature difference of 20 °C (a common wintertime temperature difference in many parts of the United States) results in an indoor-outdoor pressure difference of about –3 Pa at the bottom of the thermal stack (the basement or other ground-contact floor). Similarly, a wind speed of 4 m s^{-1} results in an indoor-outdoor pressure difference of about –2 Pa for a typical house (the relationship between the indoor-outdoor pressure difference and wind speed is quadratic, so doubling the wind speed increases the pressure difference by a factor of 4). Those are "steady-state" values. Temperature differences usually do not change very rapidly, but in the course of a day outdoor temperatures can change by 20 °C or more as part of the diurnal cycle. Wind speeds and directions are highly changeable, and this leads to substantial variation in "instantaneous" pressures. A more detailed discussion of the pressure gradients developed in buildings can be found in Liddament (1986).

The operation of mechanical systems in a building can lower the pressure in a building, especially when the flows induced by these systems are unbalanced. Operation of an exhaust fan—such as a bathroom or kitchen fan, whole-house fan, or, in some cases, an attic fan—will result in lower indoor pressures. Just as in the case of infiltration, the "makeup" air flows into the house through leaks in the building shell, some of which provide a pathway for soil-air entry. Operation of a forced-air heating and cooling system can also lead to unbalanced flows and result in lower indoor pressures, depending on the locations of the supply and return ducts and their leakage characteristics.

Diffusion

Molecular diffusion, driven by the concentration difference between low-concentration regions, such as the interior of a building, and the higher-concentration soil is another mechanism for radon entry into buildings. A key controlling variable—in addition to the radon concentration gradient—is the diffusivity of any material that separates the soil from the building interior, such as a concrete floor. In the case of an open soil floor, and in the absence of pressure differences, the radon flux density is the highest across this interface and is about the same as would be observed outdoors (soil moisture differences can have an effect on the diffusivity of soil). For soils with typical radium content, the radon flux density is $1\text{-}2 \times 10^{-2}$ Bq m^{-2} s^{-1}.

The presence of a concrete floor can increase the concentration gradient over that found in open soil, but radon diffusivities are typically smaller in concrete than for soil. The concrete floor acts as a diffusion barrier; diffusive radon entry though such a floor is likely to be somewhat lower than that for open soil. For nominal values of the diffusivity of concrete and typical radon concentrations in soil gas adjacent to a building, the radon entry rate due to diffusion through a concrete floor is about 1×10^{-2} Bq m^{-2} s^{-1}, which is about half the open-soil value. Most of this radon is from the soil itself, as opposed to the radon arising from radium in the concrete (Sextro 1994). This estimate is consistent with measurements of flux density conducted as part of extensive field experiments, where the average flux density was 1.3×10^{-2} Bq m^{-2} s^{-1} (Turk and others 1990).

Building materials themselves—especially those with soil-based constituents, such as concrete, brick, and natural stone—contain radium and will thus be a source of radon diffusion into indoor air. In most cases, however, the amount of radium in such materials is small enough that, in combination with the diffusivity of the material and typical infiltration and ventilation rates of buildings, their overall contribution to indoor radon concentrations is modest.

Other Sources

Three other sources of radon are worth noting. The first is advective transport of soil gas driven by changes in atmospheric pressure. Although large changes in atmospheric pressure can result from changes in weather, they are relatively infrequent compared with the smaller diurnal and semi-diurnal atmospheric pressure changes (Robinson and Sextro 1997). Overall, these effects are estimated to be small and to yield overall radon entry rates roughly the same as that due to the second source, infiltrating outdoor air. The latter, considered in more detail in chapter 2, provides an irreducible "baseline" indoor radon concentration.

The third source is the topic of this report: indoor use of water that contains dissolved radon, which is the subject of detailed discussions elsewhere. In the context of other sources, the average contribution made by water to indoor-air radon concentrations is very modest, given that the average transfer coefficient is 10^{-4}.

Radon Entry in Context

Several sources of indoor radon have been described; from all but outdoor air, the resulting indoor concentrations (and hence exposures) depend on the combination of the source strength and the ventilation rate of the building. As noted earlier, the stack and wind effects are primarily responsible for the natural ventilation of buildings, in addition to providing a driving force for radon transport into buildings. It is useful to provide a context for these flows.

The steady-state solution to the first-order differential equation that describes indoor radon concentrations illustrates the key variables:

$$C_i = C_o + \frac{S}{VR} \tag{8.1}$$

Where,

C_i is the indoor concentration (Bq m^{-3}),
C_o is the outdoor concentration (Bq m^{-3}),
S is the radon entry or production rate (Bq per unit time, t),
V is the house volume (m^3),
R is the removal rate (t^{-1}).

Here R can account for any method of removal and is just the sum of the individual removal terms. In this case, R is the air-exchange rate (AER). A typical single-story house has a "footprint" of 120 m^2 and a corresponding volume of about 300 m^3. Annual average natural ventilation rates are about 0.9 h^{-1}. Using those values for V and AER, the air flow rate (the product of V and AER) through this house is 270 m^3 h^{-1}.

The radon entry rate corresponding to the Environmental Protection Agency (EPA) guideline concentration of 150 Bq m^{-3} can be estimated from equation 8.1. Neglecting any contribution from outdoor air and using the ventilation rate described above, the radon entry rate, S, is about 40,000 Bq h^{-1} (about 11 Bq s^{-1}). That can be compared with the estimates of diffusive radon entry. Assuming a floor area of 120 m^2, the entry rate due to diffusion is about 1 Bq s^{-1}, a small fraction of what is needed to produce an indoor air concentration of 150 Bq m^{-3}.

Similarly, the soil-gas flow to produce this indoor concentration can be estimated. Assuming a typical value of about 40,000 Bq m^{-3} for the concentration of radon in soil gas, the soil-gas entry rate is about 1 m^3 h^{-1}, which is about 0.4% of the overall air flow rate into the house.

Mitigation Methods for Existing Houses

Conceptually, there are two approaches for mitigating indoor radon concentrations (or most other indoor pollutants, for that matter): source control and concentration reduction. One can use equation 8.1 to provide some insight into the relative efficacy of the two approaches. Indoor radon concentration is directly proportional to the source term and (again neglecting outdoor air as a source) inversely proportional to the removal terms. Considering the latter first, removal can mean either increased ventilation or some other method of removing radon or radon decay products from indoor air. In any case, for the previous example, to decrease the radon concentration by a factor of 2 by ventilation alone, the AER will need to be increased to 1.8 h^{-1}. Although that is not an excessive ventilation rate and is often achieved naturally when doors and windows are open, AER values of 2 h^{-1} commonly have comfort and energy penalties during colder seasons. Thus, this means of reducing radon concentration has some practical upper limits. In addition, forced ventilation can result in additional depressurization of a building and potentially increase the radon entry rate.

Other nonventilation removal methods are possible, and two are described in more detail below. As with ventilation, substantial removal means processing indoor air at rates that are comparable with or greater than the ventilation rate (about 270 m^3 h^{-1} in the example above). It also means that essentially the entire living space will need to be treated; this could require multiple single-room reduction devices (such as air cleaning, described below) or whole-house devices used in conjunction with a forced-air system. In the following sections, source-control methods are described first and then concentration-reduction methods.

Source Control

When high indoor radon concentrations in houses were found in various locations in North America in the middle 1970s, initial research on reduction methods was based on two key assumptions: that the source term was high concentrations of radium in soil materials derived from uranium mining and that the principal means of radon transport and entry was diffusion. Thus, initial attempts at source control focused on removal of the materials, typically uranium mill tailings used as back fill under floor slabs or adjacent to basement walls. In addition, several projects investigated the use of coatings and other sealants that would serve as an additional barrier to radon diffusion (see, for example, Culot and others 1978). Although removal of some of the high-radium-concentration materials had an effect, indoor radon concentrations in some cases were not reduced commensurately. As additional measurements of indoor radon concentration were conducted, houses were found with high indoor radon concentrations that had no known anthropogenically enhanced radon source (Sachs and others 1982). At the same time, mass-balance considerations (similar to equation 8.1 above) showed that diffusion alone had only a slight potential to produce the high indoor concentrations that were being observed (Bruno 1983).

Although removal of high-radium-concentration source materials can be part of an overall radon-control method, it generally is not part of current practice, because it is not necessary (except in rare cases). The vast majority of radon-mitigation systems now installed in existing houses rely on mechanically driven, or active, subslab depressurization (ASD) techniques (Henschel 1994). These methods seek to reverse the pressure gradient across the part of the building shell that is in contact with the soil. As noted earlier, this pressure difference drives the advective flow of radon-bearing soil gas into a building. The systems are sometimes referred to as subslab ventilation systems, but as a general rule that is a misnomer. When operated in a depressurized mode, the system does draw some outdoor air from the surface into the soil near the building. It also draws air into this region from the basement (reversing the flow of gas in the cracks and openings in the building shell). The flow of air may dilute the soil-gas radon concentration in the vicinity of the building somewhat, but the extent depends on the permeability of the soil. The key operating principle is still reversal of the indoor-outdoor pressure gradient.

Operationally, a subslab system consists of one or more pipes that penetrate the floor slab. The pipes, typically 7-15 cm in diameter, run vertically through the house and terminate above the roof. A mechanical fan, usually an in-line axial fan designed specifically for this application, is installed in the pipe system where it passes through the attic or some other location outside the conditioned living space of the house. The fan operates at about 100-400 m^3 h^{-1} at a pressure of up to a few hundred pascals (Henschel 1993).

In an ASD system, the fan creates a low-pressure zone in the soil outside the building shell. A successful system will reverse—or at least reduce—the pressure gradient at all major building-shell penetrations that are in contact with the soil. An important entry pathway for soil gas in many basement structures is the expansion-contraction joint at the edge of the concrete floor slab where it abuts the wall. In some cases, there will also be openings or utility penetrations through the basement walls or, as in the case of walls constructed of hollow-core "cinder" or concrete block, the wall itself is permeable to air flow. To eliminate or reduce soil gas entry in these areas, the low-pressure zone must extend beyond the region of the floor and up the walls. Almost all the retrofitted ASD systems are successful in reducing indoor radon concentrations to less than 150 Bq m^{-3} and often concentrations are reduced to about 75 Bq m^{-3}. In some cases when the basement walls are constructed of blocks, depressurization pipes are inserted into the hollow cores of the blocks themselves. Because these cores are typically interconnected, directly or through thin permeable concrete "webs," there is in effect a depressurized plenum within the walls themselves, thus largely eliminating any flow of soil gas across the wall and into the building interior.

Over the last decade, a considerable amount of research and practical experience on the installation of these systems has been accumulated (Henschel 1993). Two elements aid the successful implementation of an ASD system. One is the presence of a high-permeability gravel layer below the floor slab. This layer essentially establishes a low-flow-resistance pressure plenum that enhances the lateral extension of the pressure field, as described earlier. In some cases, the presence or continuity of the gravel layer cannot be easily determined. In these cases, a ~1 m diameter sump or pit is dug into the soil below the slab at the point where the ASD pipe extends below the slab. The pit helps to ensure that the pressure field created by the ASD extends as far as possible throughout the region of the soil-building interface.

The second practical element in the implementation of an ASD system is sealing as many of the potential radon-entry locations as possible. Although sealing by itself is usually not effective in eliminating radon entry, sealing does enhance the effectiveness of an ASD system because it helps to reduce any short-circuiting of air flow from the building interior into the depressurized region below the floor slab. By reducing this air leakage, the low-pressure field created by the ASD system can be further extended laterally along the soil-building interface.

One variant of the subslab system uses a fan to pressurize the region below the floor slab. In this case, the system is providing ventilation of soil gas, thus reducing the radon concentration in the soil region adjacent to the building. Rather than reducing or reversing the pressure gradient across the building shell, this method actually increases the interior-to-exterior pressure difference and so increases the flow of gas from the soil into the building. When successfully implemented, the reduction in radon concentration in the soil gas more than compensates for the increased flow. Careful studies have shown, however, that high soil permeability is key to the successful use of this technique, because it permits a larger dilution effect (Gadgil and others 1994; Turk and others 1991a; 1991b).

Basement pressurization has also been used to control radon entry. This method uses the same principle for control as does an ASD system, but it pressurizes the entire basement volume to reverse the indoor-outdoor pressure gradient. Successful use of the technique in a research-house study provided strong empirical evidence that radon entry into buildings is dominated by advective transport. However, as a practical matter, use of the technique has been limited to basements that can be made very tight with respect to air leakage, particularly the membrane between the basement and first floor. Pressurization is done with conditioned air, usually drawn from the first floor. If flow rates are too large, a substantial energy (and in some cases comfort) penalty is associated with heating or cooling the extra "make-up" air as it infiltrates into the house. This method can also create backdrafting problems for fireplaces or other combustion appliances on the first floor (Turk and others 1991a; 1991b).

The source-control methods described thus far for use in existing houses are all mechanically driven (that is, fan-powered), so-called "active" methods. Two other techniques—both passive—have been used. The first technique, sealing, has been noted earlier. Empirically, this method has not been found to reliably produce substantial reductions in radon entry, largely because it is often difficult to find or satisfactorily seal all the leakage pathways. The problem is acute when part of the soil-structure interface has a low resistance to flow, as when there is a gravel layer below the floor or when the basement walls are constructed of hollow-core block. The second passive (nonmechanical) technique that has been used as a retrofit mitigation system with some (but not uniform) success is the passive thermal stack. Similar in some respects to ASD systems, it consists of a pipe system that is inserted through the floor, and passes through the house and out through the roof. It is important that the pipe pass through the heated portion of the house, because it relies on heat transfer from this conditioned space to heat the air column inside the stack, thus creating the thermal stack effect. There is a small amount of pressure loss at each bend in the pipe, so it is also important to minimize the number of bends in the pipe system as it passes through the house.

For this system to be effective, the pressure field developed by the stack below the floor slab needs to be sufficient to reverse or at least substantially reduce, the pressure gradient between the soil and the building interior, which drives advective flow of gas from the soil into the building. The soil-to-building-interior pressure difference will be greatest when the inside-the-stack-to-outdoor temperature difference is the largest, for example, during the winter in cold or moderate climates. This is the same period when the advective transport of soil gas into the building is potentially the greatest. The influence of wind can complicate the behavior of a passive stack system. As described earlier, wind can depressurize the building interior, in addition to the depressurization caused by the stack effect. Wind can also affect the flows and pressures at the stack opening, depending on the wind direction with respect to the orientation of the roof.

It is important to have a high-permeability zone below the floor to ensure that the pressure field created by the passive stack extends along the soil-building interface, especially inasmuch as the pressure field generated by the stack is typically 1-10 Pa less than the air pressure inside the building, compared with the 100- to 400-Pa pressure difference generated by an ASD system (Gilroy and Kaschak 1990). In an existing house, the presence of such a layer and the extent to which it is present throughout can be difficult to determine.

Concentration Reduction

Unlike source-control methods, which seek to limit radon entry, concentration-control methods are designed to reduce radon or radon decay-product concentrations in indoor air. Three concentration-control techniques will be described in this section.

As mentioned earlier, increased ventilation can reduce both radon and radon decay-product concentrations, as long as it does not enhance the indoor-outdoor pressure difference. In one set of experiments conducted in a house, basement radon concentrations were observed to be lower when the basement windows were open. Measurements conducted with a tracer gas showed that basement ventilation increased somewhat with the windows open but that the largest effect was due to reducing the indoor-outdoor pressure difference across the basement wall (Cavallo and others 1996).

One method of increasing ventilation while avoiding some of the energy and comfort penalties noted earlier is to use an air-to-air heat-exchanger system, often referred to as a heat-recovery ventilation (HRV) system. In this approach, designed around commercially available HRV units, ventilation air is exhausted through a heat exchanger through which incoming unheated air also passes. The heat-exchange process is about 40-80% efficient thermally and substantially reduces the energy cost of increased ventilation (Turk and others 1991a; Fisk and Turiel 1983). Such systems have been used successfully for radon control, especially in houses with basements. Because most radon entry occurs through the basement floor and walls, basement radon concentrations are often higher than elsewhere in a house. Use of an HRV to reduce radon concentration in this space, as opposed to the whole house, means that the effective ventilation rate of the space is higher (which affords more control); by controlling basement concentrations, it also reduces radon levels throughout the house. One important ancillary benefit is that the HRV can be used to alter the basement pressure somewhat and will thus provide some additional radon-concentration reduction via source control (Turk and others 1991a).

Another method that has had very limited use is based on the sorption of radon gas by activated carbon (Bocanegra and Hopke 1989; Brisk and Turk 1984). A commercially available device based on this approach consisted of two carbon beds; one removed radon from indoor air flowing through it while the other was being purged of accumulated radon by having outdoor air passed through it and exhausted to the outdoors. The two beds were switched periodically so that the freshly purged bed was used to accumulate radon and the bed used for sorption began to be purged (Wasiolek and others 1993). Overall performance of this method is limited by the rate of air flow through the device, which in turn helps to determine the charcoal bed thickness. Like the HRV system, this approach appears to have the greatest applicability in radon control for a basement.

The third concentration-reduction approach is the use of air-cleaning to reduce radon decay-product concentrations. Unlike their chemically inert parent, the decay products ^{218}Po, ^{214}Pb, and ^{214}Bi are metals and easily attach to the surfaces of any aerosols that are present (the "attached" mode). Some decay-product atoms, particularly ^{218}Po, can also remain as ultrafine aerosols (the "unattached" mode, a few nanometers in diameter). Indoor air concentrations of both modes can be reduced by using an air cleaner designed to remove particles.

There have been a number of evaluations of air-cleaning systems undertaken in test chambers or actual indoor environments (reviewed in Hopke and others 1990). Some of these systems can effectively remove radon decay products from indoor air. However, the reduction of ^{218}Po is not as large as that of ^{214}Pb and

[214]Bi. At the same time, the particles are removed from the air. As a result, the unattached fraction of airborne activity increases, especially of [218]Po. Because the unattached fractions of the radon progeny have been considered to be far more effective in depositing their radiation dose to lung tissue, concerns have been raised regarding the efficacy of air-cleaning as a means of mitigating the hazards arising from indoor radon.

A major problem in the previous studies was that the systems used to measure radon progeny were not able to determine the full size distribution, especially in the size range below 10 nm. Estimates of the unattached fractions were made with systems that provide a poorly defined size segregation (Ramamurthi and Hopke 1989). In many cases, the size-measurement methods and results were not clearly stated.

In 1990-1992, a research program supported initially by the New Jersey Department of Environmental Protection and then also by EPA undertook field studies to investigate the effects of room-air cleaners on radon progeny concentrations and activity-weighted size distributions (Hopke and others 1993; Wasiolek and others 1993; Li and Hopke 1992; 1991b) A unique, semicontinuous graded screen-array sampling system (Ramamurthi and Hopke 1991) was used to measure the radioactivity associated with indoor aerosol particles in the size range of 0.5-500 nm.

In an early set of studies, particles were produced by a variety of activities, such as cooking, smoldering of a cigarette, burning a candle, and operating a vacuum cleaner. Aerosol behavior in the absence of an air cleaner was determined for each condition (Li and Hopke 1991a). The experiments were then repeated with a high-efficiency filter system operating (Li and Hopke 1992). It was found that the filtration unit reduced the airborne activity concentrations by removing particles, but the reductions in estimated dose were much smaller than the decrease in PAEC.

Other experiments in normally occupied houses have involved the measurement of the effectiveness of the filtration unit and an electrostatic precipitator by comparing the cumulative frequency distributions of measurements made during a week while a particular cleaner was operating and measurements made during a background week in which no cleaner was being used (Li and Hopke 1991b). A similar experimental design was used to study the two cleaners and an ionization system in an occupied home (Hopke and others 1993). The results of the 1992 measurements in Parishville, NY, in which two ionizing units were measured along with two filtration units were described by (Hopke and others 1994). More detailed studies of the NO-RAD ionizer system under the controlled conditions of a room-sized chamber at the Lawrence Berkeley National Laboratory were performed, and there are several other ionizer-based cleaners for which there have not yet been field studies (Hopke 1997; Hopke and others 1995b).

From the more recent studies on air cleaners and their effects on exposure to and dose from airborne radon decay products, several important conclusions can be drawn. With the new dosimetric models that more accurately reflect nasal and oral deposition of ultrafine particles, it is extremely unlikely that an air cleaner can reduce exposure and increase dose as suggested by Maher and others (1987), Sextro and others (1986), and Rudnick and others (1983). Thus, there is no reasonable likelihood that the use of an air cleaner will increase the hazards posed by indoor radon.

In studies of different types of air-cleaning devices, reductions in exposure have always exceeded reductions in dose. However, cases have been observed in which there has effectively been no reduction in dose. Thus, for many air cleaners, the clean-air delivery rate is insufficient to provide substantial protection from the radon decay-product hazard. The air cleaner might be effective in removing other contaminants—including cigarette smoke, dust, pollen, and spores—and thus provide a considerable benefit to an occupant without lowering the radon-progeny risk substantially and, more important, without raising that risk at all.

The experiments with the newest systems suggest that the combination of substantial air movement and ionization could provide sufficient reduction in exposure and dose to be effective in reducing the radon-progeny risk at radon concentrations up to around 400 Bq m^{-3}. If it is desirable to reduce the risk to that equivalent to the average dwelling in the United States, then such units would be useful only for lower ^{222}Rn concentrations. There would also need to be multiple units in a home to provide complete room-to-room reduction.

Mitigation Methods for New Construction

Radon-mitigation methods for new buildings can be incorporated directly into the construction process and both enhance the performance of the system and reduce the cost of installation, compared with the cost of retrofit mitigation methods. Systems for controlling radon concentrations described earlier have essentially the same applicability whether their use is in existing or new buildings, and they will not be discussed further in this section. Some cost savings might be associated with installation of systems like an HRV during the construction process or with integrating such a system into the space-conditioning system of the building.

In the following sections, the application of "existing-house" techniques for radon-entry control is discussed briefly and then a more systemic approach for making buildings radon-resistant as part of the construction process is discussed.

Application of Existing-House Radon-Entry Control Methods

As described earlier, one of the most widely used radon-mitigation techniques is ASD. Key to the successful implementation of these systems is reversal of the pressure gradient at all the major soil-gas entry points. Typically, this requires a high-permeability zone (such as gravel) below the floor and, in some cases, around the foundation footings so that the pressure field extends up the basement walls. Sealing major openings, such as at the joint of the basement floor and the wall usually is also necessary to ensure that there are few flow "short-circuits" that will degrade the pressure field. Achieving these in existing buildings can sometimes be problematic because the extent of the high permeability zone might not be known, although flow and pressure can be measured to provide a coarse assessment. In addition, it can be difficult to identify all the major soil-gas flow pathways through the building shell.

In new construction, both those problems are more readily addressed. The extent and quality of a gravel bed, for example, can be specified as part of the building design and as part of the construction-inspection process. Many of the leakage paths can be eliminated through design (for example, minimizing utility penetrations of ground-contact floors or walls), materials use (for example, use of low-shrinkage concrete for floors), and construction practices (for example, adequate sealing of utility penetrations).

One of the important benefits of these methods is that passive-stack methods might become more applicable. In some cases where wind or other effects might increase the depressurization of a house, thus potentially overriding the reverse pressure gradient established by the passive stack, the use of low-power fans for mechanical stack depressurization is attractive. Such systems have been tested in a limited number of homes and show promise (Fisk and others 1995; Saum 1991).

Radon-Resistant Buildings

Most of the elements required for making a building radon-resistant have already been described. In principle, if all entry routes through which soil gas can flow are eliminated or the pressure gradient that drives air flow through such openings is reversed, advective transport will not contribute to indoor radon concentrations. If successfully implemented, this approach can be achieved without the use of mechanical systems—it will constitute a so-called passive radon-resistance system. Such an approach has two important advantages over active radon-control systems: there are no mechanical or electric components to fail (for which the building occupants must maintain an awareness), and there is no concomitant energy use. On the other hand, the operation of mechanical systems can be easily monitored, for example, with a pressure gauge. Failure of a fan would, in principle, be easily detected by a change in pressure in the radon-mitigation pipe. The potential system-failure modes in a passive system are likely to be more subtle, such as those induced by the differential settling, cracking, and aging of building components, particularly foundation walls, footings, and floors.

Radon-resistant features, including those designed to reduce or eliminate radon-transport pathways and those in some cases, designed to reverse or decrease the differential driving pressure, have been proposed or incorporated into codes and guidelines for new buildings. In addition, some builders in various parts of the United States have voluntarily adopted construction practices that they believe will limit radon entry into new homes (Spears and Nowak 1988). In 1991, Washington state adopted a radon provision as part of the Washington State Ventilation and Indoor Air Quality Code (WSBCC 1991). It provides specific details for houses built with crawlspaces for the entire state and specifies radon-resistance features for eight counties thought to have potential for high indoor radon concentrations. The specifics include use of aggregate and a membrane below the floor slab, sealing of all floor penetrations and joints, and the use of a passive stack extending from the subslab through the heated portion of the house and exhausting through the roof. At the time of its inception, the code also required provision of a long-term radon monitor in each new home; this part of the code is no longer in force.

In March 1994, EPA published a set of standards and techniques for construction of radon-resistant residential buildings (EPA 1994a). These provisions, along with EPA's county-by-county radon zone map of the United States, were incorporated as a recommendation in the Council of American Building Officials residential building code and have been adopted, with modifications in some cases, by various local building-code authorities around the country.

In 1989, Florida initiated the Florida Radon Research Program (FRRP), which was designed to be a comprehensive program of research to examine many of the details involved in typical residential construction practice and how they might be modified to provide resistance to radon entry (Sanchez and others 1990). Many of the features are directed toward reducing radon entry in the elevated slab-on-grade construction method used widely in that state. Particular attention has been paid to attempting to ensure the integrity of the floor slab and to provide a subslab membrane that is sealed at all floor penetrations (for example, plumbing pipes) and at the edge of the slab.

The FRRP, conducted in cooperation with EPA, has produced the most extensive research on radon-resistant new construction to date. Most of the features in the proposed code have been evaluated, but in only a small number of houses (see, for example Fowler and others 1994; Hintenlang and others 1994; Najafi and others 1993; Najafi and others 1995). One part of the FRRP was the development of a radon-potential map of the state, delineating regions where no special radon controls are needed, regions where only passive (radon-resistance) features are required, and regions where both radon resistance and ASD are needed (Rogers and Nielson 1994).

Results of several house-evaluation studies conducted as part of the FRRP have shown that most of the houses built in conformity with the proposed standard appear to have short-term indoor radon concentrations below 150 Bq m^{-3}. However, there are a number of important limitations regarding these results, the most important of which might be the timing and duration of the indoor radon testing. With few exceptions, the indoor radon was measured after construction was completed but before occupancy—usually within a period of a week or two. As a result, house ventilation rates might not be similar to those during occupancy, nor will the short-term measurements of radon concentration be representative of a longer-term average indoor radon concentration.

A more generic evaluation of the effects of the different elements of the passive mitigation system proposed by the FRRP was conducted through the use of a radon-transport model with some "calibration" against data obtained in several house-evaluation projects (Nielson and others 1994). The model could not, of course, simulate failures, only the presence or absence of specified resistive features, openings through the floor, and so on. However, the results do provide insight into the relative importance of some features as applied to typical Florida residential construction. Most important was use of a vapor barrier below the floor slab, including particular attention to the details of treatment at all slab penetrations and at the slab edge, avoidance of floating slab construction, limiting concrete slump, and sealing all slab penetrations, openings, and large cracks. Interestingly, in this analysis, the presence of a passive stack to depressurize the subslab region was rated less effective than the other features of the radon-resistance system (Nielson and others 1994; Rogers and Nielson 1994). That is due in part to the assumed effectiveness of the features thought to block radon entry by reducing or eliminating air pathways between the subslab region and the house interior. In addition, the differential pressures generated across the slab by passive stacks in the houses are limited by the relatively small driving forces established by the passive stack. These occur for two main reasons: there is a narrow range of temperature differences between indoors and outdoors for most of the year, and many of the buildings are single-story (with no basement). Both limit the influence of the thermal-stack effect.

Data on the effectiveness of radon-resistance systems in other parts of the country are very sparse, and decidedly mixed. In several studies, totaling about 80 houses, measurements were made in radon-resistant houses with the passive stack closed and then after the stack was open (uncapped). Such studies appear to be based on the premise that radon-entry rates when the passive stack is closed would be similar to those observed in houses not built with radon-resistance features. However, because the passive stack is only one element in the radon-resistance system, negating its effect with a cap should not substantially affect the behavior of other parts of the radon-resistance system. More importantly, it is not clear how the radon-entry potential is affected by the use of high-permeability materials.

In one study, 46 homes were investigated in eight states (Dewey and others 1994; NAHB 1994); 41 in counties designated by EPA as having high radon potential. However, soil-gas concentrations were measured at 38 homesites (33 of them in counties designated as having high radon potential), and only 16 had soil-gas concentrations above 37,000 Bq m^{-3} (in one case the measurement was made at an adjacent site). Several researchers have suggested that soil-gas radon concentrations of at least 37,000 Bq m^{-3} (which is close to the US average soil-gas radon concentration) are needed to provide an adequate test of radon resistance in houses. This concentration was used as the selection criterion for many of the houses evaluated as part of the FRRP because it was felt that lower soil-gas radon concentrations would not test the system (Fowler and others 1994).

Of the 16 homes, four had basement radon concentrations greater than 150 Bq m^{-3}, and one of these also had radon concentrations measured on the first floor greater than 150 Bq m^{-3} (all these values were measured with the passive system fully operational, that is, with the passive stack open). In each case, these were single sets of measurements, conducted with the passive stack first closed, then open. In four of the 16 cases, the stack-open measurements were conducted in the summertime, when driving forces for both radon entry and passive-stack operation are minimal.

Other studies of radon-resistant construction have had similar results and limitations (Saum 1991; Brennan and others 1990; Saum and Osborne 1990). None of these studies had a non-radon-resistant baseline against which to evaluate the overall effect of the techniques. In many cases, soil-gas concentrations were not measured, so it is difficult to determine the radon-entry potential for the specific houses. Most of the studies conducted stack-open versus stack-closed evaluations.

Overall, the inclusion of the passive stack had the largest effect in houses with basements. In some cases, the short-term measurements suggest reductions as great as 90%, compared with stack-closed basement radon concentrations. On the average, reductions of about 40% are more typical. For slab-on-grade houses, the effect of the passive stack is considerably reduced. Comparing the stack-open and stack-closed measurements from the few houses for which there are data reveals no discernible effect. In part, that is due to the low soil-gas radon concentrations in one study. Thus, with one exception, all the initial stack-closed indoor radon concentrations are low. In the FRRP studies, soil-gas radon concentrations were often high (this was one criterion in the choice of study houses), but, as noted earlier, the passive stack was thought to be a less integral part of the radon-resistance system.

Three key questions remain unanswered:

• How well do houses built to be radon-resistant perform when compared with those built without radon-resistance features?

With one exception, no side-by-side comparisons address this question. A study of 89 homes in two areas near Colorado Springs was conducted in which 54 new homes were built within two subdivisions; 35 homes built without any radon-resistance features served as "control" homes. Of the 54 homes, 12 were tested as radon-resistant (the remaining 42 used active, mechanical systems for radon control). Radon was measured in the basements with 2-d open-face charcoal canisters, all on the same 2 days in December to remove weather effects

(Burkhart and Kladder 1991). In the original paper, the average indoor radon concentration of the 12 homes was not statistically different from that of the control homes, as analyzed either by area or in combination. Further analysis of 11 of the 12 homes examined the radon-resistance features of each home. Only five had passive stacks in addition to sealing and other techniques used to limit entry of radon; of these, only three had stacks passing through the heated portion of the homes. The three passive-stack homes averaged 130 Bq m^{-3} over the 2 days when radon was measured; one home had slightly over 150 Bq m^{-3}. The control homes averaged 450 Bq m^{-3} over the same period (Kladder and others 1991).

The only other study that has attempted to estimate the average indoor radon concentration with and without radon-resistance construction features was done as part of EPA's new-house evaluation program (Murane 1998). Measurements were made in 148 houses that were built with various radon-resistance features. All but five of the houses were built near Denver and Colorado Springs, CO and did not include a passive stack as part of radon control. The other five, built near Detroit, MI, did include a passive stack. Radon measurements were made in these houses—usually in the basement and in some cases on the first floor as well—with open-face charcoal canisters with a 2-d sampling period. The results were tabulated by ZIP code and compared with measurements in the same ZIP code done as part of the Colorado and Michigan state radon surveys (also done with short-integration time-measurement techniques). The state surveys were done in occupied houses during the winter months, but some of the new-house program measurements were done in other seasons. It is not clear whether the new houses were occupied at the time the indoor radon was measured. Overall, in the 143 Colorado houses, the average basement radon concentration was 190 Bq m^{-3}, compared with 230 Bq m^{-3} in 94 control houses in the same ZIP codes. In the case of the Detroit-area houses, there were only five radon-resistant houses and four controls; basement concentrations averaged 90 and 50 Bq m^{-3}, respectively.

Comparing the two sets of houses within each ZIP code produces considerable variation in whether the average radon concentration in the radon-resistant houses is lower or higher than of the average measured in the "control" houses. Given that all the results are based on short-term measurements, with some seasonal differences in measurements between the "control" houses and the radon-resistant ones, comparison of the average radon concentrations in the two sets of houses does not provide a strong basis for evaluating the effectiveness of the radon-resistance features. Furthermore, because one is looking for small differences between the average concentrations in the two sets—perhaps a factor of 2 or 3 at most—the study and the measurement techniques will have to be carefully designed and executed to ensure statistically meaningful results.

- Can radon-resistance systems be relied on for the lifetime of a building?

As has been indicated in the earlier discussion, there are no data on the long-term behavior of houses built with radon-resistance features. One house in Florida built and tested as part of the FRRP was revisited as part of an examination of the durability of active radon-mitigation systems. At the time of construction, provision was made for installation of an active system because the soil-gas concentrations were about 60,000 Bq m^{-3}. The postconstruction indoor radon concentration was 60 Bq m^{-3}, so an active system was never installed. The revisit, only 16 mo after construction, found essentially the same indoor radon concentration (70 Bq m^{-3}), on the basis of the results of two long-term alpha-track measurements over periods of 4 and 5 mo (Dehmel and others 1993).

- Are there limits to the applicability of purely passive radon-resistant construction practices?

Essentially no studies have explicitly examined whether there might be an upper limit to the efficacy of purely radon-resistant construction. The research conducted in the FRRP gave some indication that there could be an upper limit. For example, the postconstruction indoor radon concentration was 460 Bq m^{-3} in a house where the soil-gas radon concentration was 290,000 Bq m^{-3} (Najafi and others 1995). In another study, the author concluded that radon-resistant techniques could be used in Florida for soil-gas concentrations up to 310,000 Bq m^{-3}, as long as indoor air-exchange rates were kept above about 0.3 h^{-1} (Hintenlang and others 1994). The modeling done in support of the development of a radon potential map identified areas where the soil radon-potential was high enough that the reduction factors used for radon resistance were not sufficient to ensure that indoor radon concentrations would remain below 150 Bq m^{-3}. These regions were mapped as needing the use of an ASD system, in addition to the radon-resistance features (which often enhance the performance of active systems by limiting air flow between the interior of the house and the depressurized region below the floor slab) (Nielson and others 1994; Rogers and Nielson 1994).

In a few radon-resistant houses examined in the various studies described earlier, indoor radon concentrations exceeded 150 Bq m^{-3} (based on short-term testing). Most of these houses also had soil-gas concentrations exceeding 40,000 Bq m^{-3} (where such measurements were done). The degree to which radon-resistance construction standards and guidelines were followed was highly variable in these studies, so it is difficult to determine whether the resulting indoor (basement) radon concentrations above 150 Bq m^{-3} were due to inadequate construction techniques, low air-exchange rates (in some cases, postconstruction radon testing was done before occupancy), or inherent limits to the principle of radon-resistant construction.

Issues

All the radon or radon-progeny control systems that have been described have failure modes that can reduce or eliminate the effectiveness of mitigation.

What is more, the very nature of a successful installation of a mitigation system is to make it unobtrusive, so that without specifically thinking about it, the building occupant is not likely to know whether the system continues to perform adequately. That is true of mechanical systems—even though pressure gauges and alarms are sometimes used to signal system failure—and it is especially true of passive and radon resistant systems.

One advantage of mechanical systems is that there are objective tests of whether, for example, a fan continues to operate or, when repaired, resumes correct operation. Many ASD systems are installed with pressure gauges of various kinds that provide a necessary (but not always sufficient) measure of continued system performance. Similarly, one can determine that the fan in an HRV system or a filter system designed to remove radon decay products from indoor air continues to operate.

To the extent that radon-control systems also rely on passive radon-resistance techniques to ensure control of radon entry, failures of these features will be much harder to detect without, for example, directly measuring the indoor radon concentration. Even then, the establishment of baseline radon concentrations in a local region is necessary if one is to be able to estimate the overall effectiveness of radon-resistance construction techniques.

System Reliability and Durability

There have been only limited studies of the continued performance of active (that is, mechanically driven) radon-mitigation systems (Naismith 1997; Brodhead 1995; Dehmel and others 1993; Gadsby and Harrje 1991; Prill and others 1990). For the most part, such studies have identified two major sources of system failure: the fan ceases to operate, or is turned off by the building occupant and not restarted. One limitation to these studies has been that relatively few systems have been operating for periods approximating the mean-time-to-failure data for various fans (typically about 10 y). Buildings themselves are expected to last for many decades, if not a century, so multiple failures of a mechanical system should be expected during the lifetime of a building.

Alterations to the building, such as adding ground-contact rooms, can also alter the performance of a radon-mitigation system. In some cases, no provision for additional pipe penetrations is made so the pressure field established by the existing mitigation system does not extend into the region of the addition.

Periodic Radon Testing

One way of ensuring that systems continue to perform adequately is to conduct periodic radon or radon decay-product measurements. Typically, short-duration radon measurements are done following installation of a radon mitigation system in an existing home as a check that the system, as installed, reduces indoor radon concentrations below the EPA guideline of 150 Bq m^{-3}. Followup testing appears to be very rare. If the building is in a region of the country where radon testing is done as part of real-estate transactions, retesting might occur at that time.

In the case of new construction, particularly where radon-resistance building-construction codes are in place, there is usually no requirement for post-construction testing. Because indoor radon concentrations can be heavily influenced by the operation of a building, such as the use of a heating system (which creates the stack effect), and by occupant behavior, it is essential that radon be measured when the building is occupied. An occupant of a new home built in compliance with a radon-resistance construction code is not likely to have a strong incentive to conduct followup testing. In the case of new construction, the period between completion of construction and occupancy can be several months, and this reduces the likelihood that the construction contractor will have radon testing done.

If control of radon concentrations in indoor air is to be used as an alternative means of reducing radon (decay-product) exposures of the customers of a water-supply system, periodic testing of indoor radon concentrations will be necessary to ensure continued performance of the radon-control methods used. Because these alternatives—reduction of radon concentrations in the drinking water and reduction of radon in indoor air—can be compared only on the basis of health risks (not just indoor radon concentrations), long-term airborne-radon measurements are essential, in that they are the only basis for assessing the health risks associated with airborne radon.

Improved Estimation of the Effect of Radon-Resistant Construction

No rigorous test of the effect of radon-resistance construction practices has been done outside of work done for the FRRP. To be sure, there is evidence that radon-resistance systems can reduce the rate of radon entry into buildings in some cases, but it is not possible to determine the quantitative extent of the reduction on the basis of available data. The data are sparse, both in terms of the numbers of houses examined (for example, the variability in building types and construction practices has not been examined in detail) and in the actual followup testing procedures implemented. It is important to be able to make comparisons between similar houses built in the same area with and without radon-resistance techniques. There are a number of reasons for doing so. Aside from establishing baseline conditions with which to compare the effects of building homes radon-resistant, it will also ensure that the effects can be solidly established. Even though the current EPA building guidelines are applicable mainly in EPA zone 1 areas (the region thought to have the highest radon potential), the vast majority of homes in this region (about 86%) have annual average living-area concentrations below the 150-Bq m^{-3} guideline. That means that establishing whether radon resistance is effective will require a careful sampling design to ensure that the comparison can be established with statistical validity.

It is well known that radon concentrations in a house vary from day to day, season to season, and to some extent year to year. Some of the variations are driven by the weather and some by the behavior of the occupants indoors, such as window-opening and door-opening, and furnace and exhaust-fan operation. Thus, such comparisons should be conducted over periods long enough to average out the effects of behavior. Minimum measurement times would be two seasons, but care would have to be taken not to compare data on different houses that were taken during different seasons.

Changes due to the settling and aging of a building substructure over time can create—or extend openings through which soil gas can enter a building. Most radon-resistance approaches use a high-permeability zone just below the slab, created by either a gravel layer or a drainage mat system. The purpose of this zone is to ensure adequate lateral extension of the pressure field created by the passive stack. However, recent theoretical and experimental research has shown that such high permeability zones can substantially enhance radon entry, compared with construction in which the subslab region is not altered (Robinson and Sextro 1995; Gadgil and others 1994; Revzan and Fisk 1992). Not only might the radon-entry rate increase, because this high-permeability zone acts as a uniform-pressure plenum, but the effect of crack size (area) and location is reduced. Even cracks with small total area can transmit as much radon as openings with areas 10-15 times larger (Robinson and Sextro 1995). Thus, it is extremely important to evaluate radon resistance as buildings age to ensure that the radon resistance features are not compromised.

Research and data are needed that would permit reliable estimates of the benefits that might accrue through encouraging the use of radon-resistance construction practices in new houses as an alternative means of reducing radon-related risks. Specific research objectives are discussed in chapter 10.

MITIGATION OF RADON IN WATER

One approach to meeting the requirements of the Safe Drinking Water Act Amendments of 1996, with respect to lowering the risk associated with radon, is to treat the water directly. Several studies have evaluated water-treatment technologies for their ability to lower the radon concentration in water. Drago (1998) reported the removal efficiency, flow range, and construction cost of 34 mitigation systems now being used in small and large communities to remove radon from drinking water (table 8.1). The purpose of this section is to present an overview of existing and emerging technologies for removing radon from drinking water.

TABLE 8.1   Efficiencies, Flow, and Construction Costs for Mitigation Systems Being Used in the United States to Remove Radon from Drinking Water

Treatment Method	Removal Efficiency, %	Flow Range, $m^3 d^{-1}$	Unit Construction Cost, $ $m^{-3} d^{-1}$	No. of Systems Evaluated
I. Aeration Methods				
1. Packed tower (PTA)	79 to >99%	49 to 102,740	18 to 481	11
2. Diffused bubble				
a. Single-stage	93	431	312	1
b. Multi-stage	71 to >99	65 to 6,540	11 to 433	8
3. Spray aeration	~88[a] (estimated)	1,025	5.3	1
4. Slat tray	86 to 94	1,989 to 2,453	5.3 to 124	6
5. Cascade aeration	~88[a] (estimated)	5,450	7.9	1
6. Surface aeration	83 to 92[a]	54,504	42	1
II. Granular Activated Carbon	20 to >99	11 to 981	77 to 365	5

[a]Estimated.
Source: Drago (1998), Pontius (1998).

Aeration

In July 1991, when EPA proposed regulations for radon in drinking water, it specified aeration as the best available technology to meet the proposed maximum contaminant level (MCL) of 11,000 Bq m^{-3}. The agency's choice was based on the large removal efficiencies attainable (over 99.9%), the compatibility of aeration with other water-treatment processes, and the availability of aeration technologies in public water supplies. The documentation used to support the decision was published in 1987 (EPA 1987b) and updated in 1988 (EPA 1988a). In the 1991 proposed rule, EPA did not specify a particular type of aeration, but did cite packed-tower aeration (PTA) and diffused-bubble and spray-tower technologies. Mention was also made of less technology-intense aeration methods suitable for small water systems. Evaluations of aeration methods removing radon from drinking water are presented in Lowry and Brandow (1985), Cummins (1987) and Kinner and others (1989).

Aeration methods all exploit the principle that radon is a highly volatile gas and will readily move from water into air. The rate of removal from drinking water is governed by the ratio of the volume of air supplied per unit volume of water treated (A:W), the contact time, the available area for mass transfer, the temperature of the water and air, and the physical chemistry of radon (EPA 1987b). The dimensionless Henry's constant for radon at 20 °C and 1 atm pressure is 4.08 which is higher than values for CO_2 or trichloroethylene that are usually removed from water by aeration methods (Drago 1998).

The differences between the available aeration technologies are primarily a function of the complexity of their design and operation, the flowrates treated, and the radon removal efficiency achieved. The most efficient systems are capable of achieving >99% radon removal by increasing the surface area available for mass transfer of radon from water to air. However, these systems usually require more maintenance than simpler technologies. The more complex technologies are most practical for larger communities that must treat large volumes of water and have a large staff and tax base to support the more extensive capital and operation and maintenance requirements. Most aeration technologies require that the water system operate at atmospheric pressure to allow the release of radon to the air. This means that the systems must be repressurized to supply water to the community. Descriptions of existing and emerging aeration techniques for removing radon from water are given in appendix C.

Issues/Secondary Effects of Aeration

Intermedia Pollution

In its proposed rule, EPA (1991b) recognized that emissions from aeration systems potentially could result in a degradation of air quality and pose some incremental health risk to the general population because of the release of radon to the air. On the basis of EPA's analysis (EPA 1989; EPA 1988b), the increased risk is much smaller than the risk posed by radon in the water. In its initial evaluation with the AIRDOSE model, EPA (1988b) used radon concentrations in water of 68,000 Bq m^{-3} (range, 37,000-598,000 Bq m^{-3}) based on data from 20 water systems in the United States. Assuming a 100% transfer of radon from the water to air, EPA estimated that radon would be emitted into the ambient air at 0.10 Bq y^{-1}. EPA used an air dispersion model (including radon and its progeny) and assumed ingestion and inhalation exposures in a 50-km radius, and it calculated a maximal lifetime individual risk of 4×10^{-5} (0.016 cancer case y^{-1}). Extrapolated to drinking-water plants throughout the United States, that translated to 0.4 and 0.9 cancer cases per year due to off-gas emissions from all drinking-water supplies meeting MCLs of radon of 7,400 or 37,000 Bq m^{-3} in water, respectively. EPA used a similar approach to assess the risks associated with dispersion of coal and oil combustion products.

In an evaluation with the MINEDOSE model, EPA (1989) used worst-case scenarios from four treatment facilities whose raw water radon concentrations were 49,000 to 4,074,000 Bq m^{-3}. In only one facility was there a significant potential increase in cancer risk associated with radon emissions when a single point source was assumed. However, EPA found that this large water utility actually used a number of wells at various locations, instead of one source, and that reduced the risk because of dispersion over a greater area. The highest raw water radon concentrations did not always result in the greatest effect on air quality because high concentrations often occurred in small systems with low flow rates, which yielded lower overall emissions. The evaluation concluded that the resulting health risk posed by radon release into the atmosphere via aeration-system off-gas was much smaller (by a factor of about 100 to 10,000) than the risks that would result if radon were not removed from the water.

The EPA Science Advisory Board (SAB) reviewed EPA's report (EPA 1989; EPA 1988b) and found that the uncertainty analysis needed to be upgraded to lend more scientific credibility to the air-emissions risk assessment. However, the SAB also stated that revisions in the modeling would not change EPA's conclusion that the risk posed by release of radon from a water-treatment facility would be no more than the risk posed by using drinking water that contains radon at 11,000 Bq m^{-3}. The SAB also noted that EPA's assumptions were conservative.

EPA also had its Office of Radiation and Indoor Air (ORIA) review its 1988 (EPA 1988b) and 1989 (EPA 1989) air-emissions studies for consistency and to provide a simple quantitative uncertainty analysis (EPA 1994b). The ORIA review indicated that the early studies had overstated the risk; it estimated an incidence of cancer of 0.004 cases per year, less than the 0.016 case per year initially estimated.

EPA (1991b) acknowledged in the proposed rule for radon that "some states allow no emissions from PTA systems, regardless of the downwind risks." Indeed, on the federal level, EPA has, under the Clean Air Act, established National Emission Standards for Hazardous Air Pollutants (NESHAPs), including radon. Under NESHAP, an average radon-emission rate of 0.74 Bq m^{-2} s^{-1} is allowed from radium-containing facilities, and an individual member of the public can have a maximal exposure of 1.00×10^{-4} Gy y^{-1}. It is not clear whether NESHAPs will ever affect aeration systems, inasmuch as they are applicable only to industries, not to drinking-water treatment facilities. Drago (1998) has noted that if the radon NESHAP were applied to a 93 m^3 d^{-1} water-treatment plant (serving about 250 people) with an influent radon concentration of 18,500 Bq m^{-3} in its water, radon would be emitted in the off-gas at 222 Bq m^{-2} s^{-1}. This analysis suggests that if the NESHAP were extended to include drinking-water plants, many aeration systems would have to treat the off-gas to remove radon. The Nuclear Regulatory Commission also regulates radon releases from the facilities that it licenses, but it is doubtful that its limits would ever apply to water-treatment facilities. Some states have adopted NESHAPs for radionuclides or Nuclear Regulatory Commission limits by agreement, but none directly applies them to water-treatment plants.

It is possible that some states may conduct or require risk screening of new water treatment aeration facilities for radon emissions. Although some states allow specific incremental lifetime risks associated with hazardous air pollutants (for example, California one in 1 million), it is not clear that they will be applied to radon. Drago (1998) noted that only Nevada, California, New Jersey, and

Pennsylvania have included radon in risk assessments of drinkingwater aeration facilities and that in these cases, radon was part of an evaluation in which volatile organic compounds (VOCs) were the contaminants being regulated, not radon itself.

If radon had to be removed from the off-gas leaving an aeration system, it would present a problem because no efficient and cost-effective technology is available. Granular activated carbon (GAC) is often proposed for off-gas treatment, but as EPA (1991b) acknowledged in its proposed rule, it would probably not be effective in removing radon from air. GAC can remove vapor-phase radon, but it is relatively poorly adsorbed and, according to the equation derived by Strong and Levins (1978), the empty-bed contact time (EBCT) required would be many hours. Martins and Meyers (1993) estimated that less than 2% of radon would be removed by typical vapor-phase GAC units currently used at water-treatment plants to remove VOCs (where EBCT are in seconds). Only with complex methods of concentrating the radon, perhaps by adsorption and desorption (Thomas 1973) and the use of several GAC beds in series could sufficient removal be achieved. Even then, performance would be difficult to control because vapor-phase adsorption is a function of temperature, humidity, and interferences by other gaseous compounds (Bocanegra and Hopke 1987). Indeed, very few studies of techniques to remove radon from off-gas (vapor-phase removal) have been conducted, even in the nuclear industry. Drago evaluated the work of Bendixsen and Buckham (1973) on removing noble gases from gas streams at nuclear facilities (table 8.2.) and found that the methods available are impractical or cost-prohibitive for use in the water industry. As a result, EPA (1991b) suggested in its proposed rule that at sites where air emissions regulations prevent release of radon from aeration systems, GAC might need to be used, instead of aeration, to treat the water. Some other technologies, not known in 1991, when the proposed rule was published, might also be alternatives to aeration, but these are only in the testing phase (for example, vacuum deaeration). The issue of

TABLE 8.2   Methods to Remove Noble Gases from Gas Streams at Nuclear Facilities

Adsorption
    in fluorocarbons[a]
    in CO_2[a]
    on activated carbon[a]
Cryogenic distillation[a]
Membrane separation[a]
Cryogenic oxidation with fluorinating compound[b]
Cryogenic membrane separation[b]

[a]Bendixsen and Buckham (1973).
[b]Drago (1998).

radon removal from aeration-system off-gas remains unresolved and would probably be of greatest concern in urban areas.

Airborne release at facilities that treat groundwater can expose operators to high concentrations of radon (Fisher and others 1996). Ironically, the problem has been identified at plants that treat groundwater for contaminants other than radon (such as iron). In a survey of 31 water-treatment plants in Iowa, Fisher and others (1996) found that processes such as filtration, backwashing, and regeneration cause radon release directly into the plant. Their results suggest that the air in all facilities that treat groundwater should be monitored for radon and that ventilation should be investigated as a means of reducing worker exposure.

Microbial and Disinfection-Byproducts Risks

Treated water that leaves aeration systems might contain increased bacterial counts (Kinner and others 1990; 1989). On the basis of the pending Groundwater Disinfection Rule (GWDR) (EPA 1992b), disinfection would be required in cases where the heterotrophic plate count exceeded 500 colony-forming units per milliliter. In addition, aeration systems can have periodic problems with high coliform counts in the treated water as a result of the transfer of bacteria from air to water during treatment. That might also necessitate disinfection to comply with the coliform rule in distribution systems (Drago 1998).

EPA did not mention the potential need for disinfection of the effluent from aeration systems in its 1991 proposed rule, nor did it consider disinfection in its cost estimates for radon treatment. The SAB (1993) criticized the agency for neglecting to do so, and EPA has since added these costs (EPA 1994b). The technology to disinfect groundwater is well developed, and disinfection systems already exist in some communities or are being added to meet the requirements of the pending GWDR. The commonly used disinfection methods include chlorination (such as with sodium hypochlorite or gaseous chlorine) and ultraviolet irradiation.

Although disinfection reduces the health risk resulting from microbial contamination of drinking water, it has its own associated risks, especially if chlorination is used. Groundwater might contain organic carbon at 0.5 to 2 mg L^{-1} (Cornwell and others 1999; Miller and others 1990; Kinner and others 1990; 1989). Addition of chlorine to water that contains such natural organic matter could result in the formation of disinfection byproducts (DBPs) (that is, trihalomethanes, THMs—such as chloroform, dibromochloromethane, dichlorobromomethane, and bromoform—and other compounds at lower concentrations) in the range of 10-50 µg L^{-1}. THMs are regulated in water under the Disinfection By-Products Rule because they are known to cause cancer in rats and mice. As a result, EPA has established potency values for these compounds. Disinfecting the effluent of aeration systems that remove radon from water increases the risk of exposure to disinfection byproducts. The SAB (1993) criticized EPA for not discussing the cancer risks associated with exposure to disinfection byproducts; to date, the agency has not done so.

According to Wallace (1997), finished water produced from surface water tends to have higher THM concentrations than finished water from groundwater supplies. One of the concerns of this committee with regard to methods of reducing radon exposure is the potential for increased exposure to THMs if radon mitigation results in the use of chlorine to disinfect the water to satisfy the pending GWDR. To examine this issue, the committee made a screening level estimate of the relative change in cancer risk associated with surface water and groundwater.

Chloroform concentrations are monitored extensively at water-treatment plants, but only sporadically in residential tapwater. Wallace (1997) has reviewed a number of surveys of water-supply concentrations of THMs. Among them, the Community Water Supply Survey provides representative and comprehensive results (Brass and others 1981). The (average and median) values of chloroform, dibromochloromethane, dichlorobromomethane, and bromoform in supplies derived from surface water were 90 (60), 12 (6.8), 5 (1.5), and 2.1 (<1) µg L^{-1}; respectively, and the average values of chloroform, dibromochloromethane, dichlorobromomethane, and bromoform in supplies derived from groundwater were 8.9, 5.8, 6.6, and 11 µg L^{-1}; respectively, average concentrations all below the detection limit.

The committee used those reported concentrations with unit dose factors for inhalation, ingestion, and dermal uptake and with EPA cancer potencies of the compounds to make approximate risk estimates. The average lifetime cancer risk associated with a surface-water system is around 1×10^{-4} and the corresponding risk associated with a groundwater system is around 5×10^{-5}, smaller than the 1×10^{-4} risk of lung cancer posed by inhalation of radon released from water (see chapter 5). The calculations are summarized in appendix D.

Corrosion

Aeration during radon treatment increases the pH of water (Kinner and others 1990; 1989). The increase has been attributed to the removal of CO_2 from the water. In a study of aeration units used for VOC treatment, the American Water Works Association (AWWA 1991) reported that the effect of CO_2 removal, with the greater stability of $CaCO_3$ at the higher pH, negated the effect of the increased oxygen concentration in water. There was no increase in the corrosivity of the water. At one very small water-supply system in Colorado, aeration of the water to remove radon actually eliminated the need for addition of lime to prevent corrosion (Tamburini and Habenicht 1992). At a small system in New Hampshire, aeration resulted in a decrease in corrosivity and a reduction in the lead and copper measured in the drinking water (personal communication, D. Chase, Department of Health and Human Services, Bureau of Radiological Health, August

15, 1997). Although those studies suggest that corrosivity decreases with aeration, it might still be necessary in some systems to add corrosion inhibitors (such as lime and sodium silicate) to reduce the potential for increased release of lead and copper from the plumbing and distribution system.

Precipitate Formations

One problem observed in some aeration systems is formation of precipitates (scale) which can cause operational problems (such as fouling of equipment) and aesthetic concerns (such as release of precipitates to consumers). Depending on the chemical characteristics of the raw water, the most common precipitates are oxide, hydroxide and carbonate species of iron, manganese and calcium. Typical A:W ratios for radon removal systems might be 15:1, and this could result in precipitate formation at iron concentrations as low as 0.3 mg L^{-1} (Kinner and others 1993).

Common methods of eliminating precipitate-formation problems involve periodic addition of weak acid solutions to clean the equipment or addition of sequestering agents that bind the cations (Dyksen and others 1995; AWWA 1991). Another approach is to install cation-exchange filters before the aeration system. These filters are very effective at trapping iron, manganese, or calcium, but they also concentrate other naturally occurring cations, some of which can be radionuclides (such as Ra^{2+}). The brine used to regenerate the ion-exchange filters can also become contaminated with long-lived radionuclides. In some small-scale applications, aeration equipment is followed by sand or cartridge filters that trap the precipitates (Drago 1998; Malley and others 1993). That is a simple method of removal especially useful in plants that do not have full-time professional operators. After sufficient precipitate has collected in the filter, substantially decreasing water flow, it must be backwashed. Both the brine from the ion-exchange unit and the backwashed material from the sand or cartridge filter are usually discharged to the nearest sewage-treatment system, which in many rural areas is a subsurface leach field. In its cost estimates for radon treatment with aeration, EPA did not adequately consider the cost of precipitate treatment, nor did it address adequately the problems of precipitate formation.

In some cases, it has been shown that iron precipitates have enriched concentrations of long-lived radionuclides, such as radium, lead, or uranium (Cornwell and others 1999; Kinner and others 1990). Depending on the concentrations of these radionuclides, the brine or backwash residuals might require special disposal, as discussed in EPA's (1994c) *Suggested Guidelines for Disposal of Drinking Water Treatment Wastes Containing Radioactivity*. Special treatment of the residuals can increase the cost of operation and the risk to workers who must oversee the disposal process.

Granular Activated Carbon

In the 1991 proposed rule for radionuclides in drinking water, EPA stated that GAC was not a best available technology for radon removal, although it has been shown to remove radon from drinking water (Kinner and others 1989; Lowry and Lowry 1987; Lowry and Brandow 1985). The agency cited problems with radiation buildup, waste disposal, and contact time. Since then, the SAB (July 1993, EPA-SAB-RAC-93-014 and July 1993, EPA-SAB-DWC-93-015) (EPA-SAB 1993b) has suggested that GAC might be an option for small systems with modest raw-water radon concentrations and that there could be problems with the thoroughness of EPA's analysis of the risk and disposal issues related to the use of GAC. In addition, new data that have become available since 1991 suggest that GAC might require shorter empty-bed contact times than originally thought (Cornwell and others 1999).

GAC was first shown to be an effective technique for removing radon from drinking water in the early 1980s (Lowry and Brandow 1981). As a result of its simplicity, it was installed in many private homes in New England where radon levels were high (1,111,000 to 11,111,000 Bq m^{-3}) (Lowry and others 1991). By the late 1980s, studies of GAC units were producing data that suggested that gamma emissions from ^{214}Bi and ^{214}Pb, the short-lived progeny of radon, were substantial ($2 \times 10^{-6} - 5 \times 10^{-4}$ Gy h^{-1}) (Kinner and others 1989; Lowry and others 1988; Kinner and others 1987). In addition, accumulation of long-lived species (such as uranium, radium, and especially ^{210}Pb) on the GAC was creating disposal problems (Kinner and others 1990; Kinner and others 1989; Lowry and others 1988). The radon-removal efficiency of some GAC units also decreased with time (Kinner and others 1993; Lowry and others 1991; Kinner and others 1990; 1989). It was the data from the studies in the late 1980s that led EPA to question the use of GAC as a best available technology in its 1991 proposed rule. Furthermore, economic evaluations suggested that the cost of GAC treatment is high and in most situations not competitive with aeration, because of the large amount of carbon needed, especially for large radon loadings (high flow or high influent radon concentration) (EPA 1987b). A description of the GAC process is found in appendix C.

Retention of Radionuclides on GAC

By their very nature, GAC systems are designed to sorb and retain contaminants. Thus, while a GAC unit is operating, the bed is accumulating radon. In addition, because of radon's short half-life relative to the GAC unit's run-time (months to years), radon comes to secular equilibrium with its progeny. The solid progeny remain sorbed to the GAC (Cornwell and others 1999; Kinner and others 1990; Kinner and others 1989; Lowry and Lowry 1987) which is not surprising inasmuch as GAC is known to have a high affinity for metals (such as lead) relative to the small mass of them created from the decay of radon (Reed and Arunachalam 1994; Rubin and Mercer 1981) (for example, GAC can sorb lead at $6.2 \times 10^4 - 1.9 \times 10^6$ µg kg^{-1} at a pH of 6.5).

Some groundwater also contains radium and uranium in addition to radon. Uranium can sorb directly to the GAC, but its fate is a function of the pH of the water. At a pH greater than 7, the poorly sorbed, negatively charged carbonate species of uranium, $UO_2 (CO_3)_2^{-2}$, is predominant. At a pH lower than 6.8, the neutral species, UO_2CO_3, is predominant and can sorb to GAC (Sorg 1988). Radium is poorly sorbed to GAC (Kinner and others 1990; Clifford 1990; Kinner and others 1989; Sorg and Logsdon 1978) because it forms a hydrophilic species, $RaSO_4$, in water. The pattern and rate of accumulation of uranium, radium, and ^{210}Pb in a GAC unit can be quite different if iron is present. Cornwell and others (1999) found high concentrations of these radionuclides associated with iron-rich backwash residuals from GAC units. That is because radium readily associates with ferric hydroxide and negatively charged metal oxides and hydroxides (Clifford 1990). Uranium also reacts with iron (Clifford 1990; Sorg 1988).

Operational Issues.   During operation of a GAC unit, an equilibrium is established between the radioactivity of radon and its short-lived progeny sorbed to the carbon. The primary problem resulting from retention of radionuclides is worker exposure to gamma emissions from ^{214}Bi and ^{214}Pb. The maximum occupational accumulated dose equivalent per year recommended for radiation workers in the United States is 50 mSv (EPA 1987a). However, EPA has stated that "there is no need to allow" workers in water-treatment facilities that remove naturally occurring radionuclides from water to receive such high annual radiation doses. It further suggests that these workers' annual accumulated dose equivalent should be "well within the levels recommended for the general public" of 1,000 µSv. Hence, EPA has recommended a maximum annual administrative control level of 1,000 µSv until more experience with such situations is gained.

Lowry and others (1988) measured the gamma-exposure fields surrounding 10 point-of-entry units treating water with radon at 96 to 28,074,000 Bq m^{-3} and achieving removal efficiencies of 83% to over 99%. The gamma exposure rates measured at about 1 m were considerable, in all but one case, because the radon concentration removal was very large (C_{net} = 611,000 to 27,926,000 Bq m^{-3}). Except for the 27,926,000-Bq m^{-3} case, the measurements are in agreement with the range of the calculations from the extended source model.

The gamma exposure to workers can be decreased by using water or lead shields around the GAC units. Lowry and others (1991; 1988) studied the effect of water shielding and lead jackets on the point-of-entry units' gamma exposure fields. For example, at site 9 (table 8.3) (28,074,000 Bq m^{-3}) the maximum gamma-exposure rate at the unit's surface was 73 mR h^{-1}. With a 76.2-cm water shield, this was reduced to 8.0 mR h^{-1}. A 61.0-cm water shield reduced a maximum surface gamma-exposure rate of 4.0 mR h^{-1} to 0.4 mR h^{-1} at site 5; and at

TABLE 8.3   Maximum Gamma-Exposure Rates and Equivalent Dose Rates from Some Point-of-Entry GAC Units at a Distance of 1 m[a]

Site	Average Radon $C_o{}^b$ (Bq m^{-3})	Average Radon $C_t{}^c$ (Bq m^{-3})	Exposure Rate (mR hr^{-1})	Equivalent Dose[d] (µSv hr^{-1})
1	97,000	14,000	NA[e]	NA[e]
2	613,000	2,200	0.40	4.0
5	1,989,000	165,000	0.186	1.86
5 (with 61-cm water shield)	1,989,000	165,000	0.040	0.40
9	28,104,000	175,000	1.73	17.3

[a]Lowry and others (1988).
[b]Input water concentration.
[c] Treated water concentration.
[d]Equivalent doses were calculated by the committee and were not included in Lowry and others (1988). See appendix E for method of calculation.
[e]NA = gamma radiation so low that it was impossible to measure with acceptable accuracy.

about 1 m, there was also a significant reduction. Lead shielding (0.64 cm thick) also reduced gamma exposure.

It is possible to estimate the equivalent dose in millisievert per hour from a GAC unit that removes radon from water if an extended-source model is used (see appendix E). This type of model calculates radiation doses in the vicinity of an extended source of radioactive material including self-shielding. This approach is used in the nuclear industry and in radiation protection to predict the radiation levels associated with a variety of radioactive materials (such as ion exchange units that treat nuclear-reactor cooling water and steam condensate).

GAC units operating in the United States are treating 11-981 m^3 of water per day (table 8.1). GAC is also used in point-of-entry applications (water flowrate of 1 m^3 d^{-1}). For the purposes of calculating the equivalent dose with the extended source model, the committee used the suggestion made by Rydell and others (1989) that GAC should be used only to treat water that contains radon at less than 185,000 Bq m^{-3}. The unit would be required to meet the MCL, which is assumed to be 25,000 Bq m^{-3} for this example. (The committee makes no recommendation or endorsement of a specific value for the radon MCL and uses 25,000 Bq m^{-3} in order to provide a framework for the example.) The results (table 8.4) indicate that as the radon loading (becquerel applied per day) increases with increasing water flow rate over the range of 1 to 981 m^3 d^{-1}, the time until the 1,000 µSv maximum equivalent dose is reached decreases from about 7,000 h to about 150 h (1 m from the GAC tank surface). In many cases, it is unlikely that water-treatment plant personnel would need to spend hundreds of hours per year near the GAC units. In fact, the actual number of hours of exposure per worker would be different for each water supply. Certainly, the number of hours of

TABLE 8.4   Estimated Equivalent Gamma Dose for Workers at Water-Treatment Plants or in Point-of-Entry Applications Using GAC[a]

Water Flow (m^3 d^{-1})	Estimated Gamma Dose at 1 m (μSv hr^{-1})	Hours until 1,000 μSv Dose
1 (point of entry)	0.14	7,143
11 (water plant)	0.75	1,333
981 (water plant, pressure driven)	7.0	143
981 (water plant, gravity driven)	6.4	156

[a]See appendix E for calculations.

exposure would need to be monitored to ensure worker safety. If it were necessary to work on a GAC unit for a substantial number of hours, the gamma emissions could be reduced by first taking the unit off line for about 20 to 30 d to allow the radon responsible for the short-lived progeny to decay.

The calculations also show that as radon loading increases, the dimensions of the tank increase, providing increased absorption of gamma radiation within the tank. Modeling gamma emissions from the tank as a point source will be satisfactory only for very small units (for example, point-of-entry applications). The equivalent gamma dose from a GAC system that removes radon from a public water supply should be modeled with an extended-source model that can be modified to the dimensions of the treatment units. It is clear that treating water that contains more radon (over 185,000 Bq m^{-3}), where high removal efficiencies are required, or at high flow rates (high radon loading) will probably lead to unacceptable equivalent gamma doses to water-treatment plant personnel.

Rydell and Keene (1993) have developed a computer program (CARBDOSE 3.0) that calculates the probable gamma-exposure dose with distance from a typical point-of-entry unit. The program "approximates a 25.4 cm diameter, 12.7 cm high cylindrical volume of GAC as a cylindrically-corrected 24 cm × 24 cm × 13 cm array of 1 cm^3 sources using the 72 gamma energies reported for [214]Bi and [214]Pb and allowing for self absorption and build-up." Rydell and others (1989) reported that the CARBDOSE models' estimated gamma dose rates and the measured values for 10 point-of-entry GAC units were in "reasonably good agreement." They suggest that CARBDOSE can be used as a design tool to estimate the potential gamma radiation exposure during operation of the GAC unit. The committee notes that CARBDOSE should only be applied to GAC units that have very small dimensions (that is, ones that treat very small flows) and are similar to those used in developing the model. Equivalent gamma doses for larger GAC units should be predicted with an extended-source model that can address more-complex geometries.

Disposal Issues.   A few weeks after a GAC unit ceases operation, the major radionuclide remaining sorbed to the carbon is ^{210}Pb because of its relatively long half-life (22.3 y). In cases where iron is associated with the GAC or the raw water has a pH less than 6.8, uranium and radium can also be found. These species can pose problems for the long-term disposal of the GAC.

No federal agency currently has legislative authority concerning the disposal of drinking-water treatment-plant residuals that contain naturally occurring radionuclides (Cornwell and others 1999). If the GAC is transported to a site for disposal, the Department of Transportation could regulate its shipment. EPA has published two guidelines that suggest how such wastes might be handled (EPA 1994c; 1990). However, the states are responsible for regulation of naturally occurring radioactive materials (NORM). There have been three detailed reviews of federal and state guidelines and regulations regarding NORM and how they might apply to disposal of GAC used to remove radon from water (Cornwell and others 1999; Drago 1998; McTigue and Cornwell 1994).

The EPA (1994c) guidelines for disposal of water-treatment residuals are centered on the levels of uranium and radium present (for example, in spent GAC or backwash residuals) (table 8.5). Unlike its 1990 draft guidelines, EPA's 1994 version did not cite specific action levels for ^{210}Pb. Instead, because of a lack of conclusive technical data, EPA recommended that the impact of ^{210}Pb contamination be considered case by case. Most states also address NORM wastes on a case by case basis (Drago 1998); the exceptions are Illinois, Wisconsin, and New Hampshire, which have established disposal criteria (Cornwell and others 1999).

The Conference of Radiation Control Program Directors published a draft set of suggested state regulations for technologically enhanced NORM (TENORM), naturally occurring radionuclides whose concentrations have been enhanced by technology (for example by such practices as water treatment). Lead-210 associated with GAC is not specifically addressed in this document, and materials with ^{226}Ra or ^{228}Ra at less than 0.19 Bq g^{-1} are exempt. The draft recommends flexibility

TABLE 8.5   EPA Suggested Guidelines for Disposal of Naturally Occurring Radionuclides Associated with Drinking-Water Treatment Residuals

Radionuclide	Bq g^{-1} (dry weight)	Suggested Disposal Site
Radium	<0.11	Landfill
	0.11-1.85	Covered landfill
	1.85-74	Possible RCRA facility (case by case review)
	>74	Low-level radioactive-waste facility
Uranium	<1.11	Landfill
	1.11-2.78	Covered landfill
	2.78-27.8	Possible RCRA facility (case by case review)
	>27.8	Low-level radioactive-waste facility
^{210}Pb	—	Caution and thorough state-agency review of water treatment and waste disposal plans

Source: EPA (1994c).

in regulating TENORM as long as members of the public receive less than 1×10^{-3} Gy y^{-1} from all licensed sources (including TENORM).

If spent GAC from a water-treatment plant had enough ^{210}Pb, radium, or uranium associated with it to warrant disposal at either a low-level radioactive waste site or a naturally occurring and accelerator produced radioactive materials (NARM) site, this could have a substantial impact on operation and maintenance costs for the water utility. Actual disposal costs have been estimated as $335 m^{-3} yr^{-1} (Kinner and others 1989), approximately $48,000 m^{-3} (McTigue and Cornwell 1994) and about $11,100 m^{-3} y^{-1} (Cornwell and others 1999). In addition, broker and transportation fees would likely be assessed. A typical broker would send trained personnel to the treatment plant to dewater the bed, load and seal the GAC in containers, and decontaminate the site. Cornwell and others (1999) estimated the broker fee at $5,000 (mostly associated with time and travel).

Perhaps the biggest question surrounding GAC disposal is the availability of sites that will accept such radioactive material. Drago (1998) reported that two sites are operating (in Barnwell, SC, and Richland, WA). (Note: Clive, UT, receives only limited low-level and NARM wastes.) However, these facilities are not available to all states. Rather, the Low Level Radioactive Waste Disposal Policy Act (PL-99-240) enacted in 1980 and its amendments (1985) direct states to form compacts with their neighbors and designate a host low-level disposal site. There are nine compacts and one other pending, and five states, Washington, DC, and Puerto Rico are unaffiliated. Low-level disposal sites have been proposed by some of these compacts, but none has been built. The result is that low-level waste generators in all states except North Carolina have access to a disposal facility (Drago 1998), but new facilities are not likely to be readily available in the near future.

Several ways of avoiding the need to dispose of the GAC at a low-level waste facility would not require changing legislation or regulations. Perhaps the easiest would be to dispose of the GAC before radionuclide accumulation necessitates special disposal. McTigue and Cornwell (1994) developed a model that allows operators to predict when a bed is reaching such a level with respect to [210]Pb. The CARBDOSE model (Rydell and Keene 1993) makes a similar prediction for POE GAC units. These models are simple to use, and periodic measurements of the actual ^{210}Pb accumulation on the GAC can be made to confirm their estimates. It should be noted that the models do not address the effect of GAC-associated-iron on the ^{210}Pb accumulation (Cornwell and others 1999). If substantial amounts of iron were present in the raw water, such a prediction would be more difficult.

Another alternative to disposal of the spent carbon is thermal regeneration of the GAC that Lowry and others (1990) showed was possible. Both ^{210}Pb and its progeny are volatilized at 850 °C. It is not clear whether release of the ^{210}Pb or [210]Po to the atmosphere would be acceptable. If those radionuclides were collected in an air scrubber, they would potentially still present a radioactive-waste disposal problem with respect to the fly ash. Acid regeneration of the spent GAC is also possible (Lowry and others 1990). In this case, ^{210}Pb, like stable lead (Reed and Arunachalam 1994), would desorb from the GAC and enter the acid-regenerant solution. However, the spent acid could become a radioactive waste that requires special disposal.

Several authors (Cornwell and others 1999; McTigue and Cornwell 1994) explored the possibility of the GAC's being returned to the vendor (an approach used for GAC used to treat VOCs or substances that impart taste and odor to water). However, the willingness of the manufacturers to do this with radioactively contaminated GAC is not clear, especially for small quantities of GAC (less than 9,100 kg).

The best option overall with respect to disposal appears to be use of GAC in sites where the potential for ^{210}Pb accumulation is minimized (that is, where the radon and iron concentrations in the raw water are low or the water flow rate is low. This would ensure fairly long operating times before the ^{210}Pb reached a critical level likely to necessitate special disposal. The low radon loading would also result in lower risk of worker exposure to gamma radiation.

Long EBCT and High Cost

Bench-, pilot- and full-scale studies of GAC removal of radon have produced estimates of the K_{ss} (adsorption-decay constant, see appendix C) for different carbons (Cornwell and others 1999; Kinner and others 1993; Lowry and others 1991; Lowry and Lowry 1987). Lowry and Lowry (1987) found that the best carbon for radon sorption was a coconut-based GAC ($K_{ss} = 3.02$ h^{-1}). This carbon has a larger percentage of micropores (0.002 μm) than other types of GAC. It is hypothesized that micropores are most effective for sorbing small molecules and atoms, such as radon gas (Drago 1998).

The cost estimates for GAC treatment have used a K_{ss} of less than 3.02 h^{-1} (for example, EPA 1987b, $K_{ss} = 2.09$ h^{-1}). Recent studies by Cornwell and others (1999), specifically designed to calculate K_{ss} values for different carbons, found that for one groundwater with low iron and TOC concentrations, the K_{ss} ranged from 3.5 to 5.2 h^{-1}. These higher values suggest that GAC could be a much more cost-effective option at some sites than originally thought. For example, with a raw-water radon concentration of 111,000 Bq m^{-3}, a flow of 39 m^3 d^{-1}, and a K_{ss} of 4.5 h^{-1}, the EBCT and amount of GAC needed to achieve an MCL of 11,000 Bq m^{-3} (90% removal) would be 31 min and 0.83 m^3, respectively, compared with 66 min and 1.8 m^3, respectively, if the K_{ss} were 2.09 h^{-1}. Assuming a cost of $883 m^{-3} of GAC (Cornwell and others 1999), this 54% reduction translates to an $848 savings. However, Cornwell and others (1999) also suggest that a pilot study must be conducted at each site where GAC is being considered, because the K_{ss} will likely be different for each groundwater. For example, for a low-iron, low-TOC groundwater in New Hampshire, a lignite-coal based GAC had an average K_{ss} of 4.52 h^{-1}. At a site in New Jersey that also had low iron and low TOC, the average K_{ss} for the same carbon was 2.54 h^{-1}.

Microbial Risk

GAC units often support microbial populations because they provide a good surface for attachment and concentration of organic carbon and nutrients (Camper and others 1987; Graese and others 1987; Camper and others 1986; 1985; Wilcox and others 1983). As a result, GAC units that treat radon have had high concentrations of heterotrophic bacteria in their effluent (Cornwell and others 1999; Kinner and others 1990; 1989). Because heterotrophic plate counts could periodically exceed 500 colony forming units per milliliter, which would violate the proposed GWDR, the treated water would need to be disinfected before distribution (EPA 1992c). Either chlorination or ultraviolet disinfection could be used. Unlike the situation with aeration systems, it is less likely that disinfection byproducts would be formed if chlorination were used after a GAC unit. Disinfection byproducts result from the reaction between chlorine-based disinfectants and naturally occurring organic matter in the water. GAC can sorb the low levels of naturally occurring organic matter in the groundwater (Cornwell and others 1999) until it becomes saturated (Kinner and others 1990). Before saturation, the risk posed by disinfection byproducts would be minimal. Thereafter, it could be similar to that of an aeration system that uses chlorine-based disinfection.

Precipitate Formation

In groundwaters that have high levels of iron, precipitates might accumulate on the top of the GAC bed. This reduces the hydraulic head and contaminant-removal efficiency of the GAC and makes it a poor choice for radon treatment for these types of raw water. Although some iron precipitation has been observed in field evaluations of GAC units that treat radon with low iron (Cornwell and others 1999; Kinner and others 1990; 1989), the problem is usually less acute than in aeration treatment; the water is not usually oxygenated and exposed to atmospheric conditions, so, much less oxidation of the iron occurs. If iron precipitates do form, they present the same problems outlined for aeration systems. Pretreatment to remove the iron before the water enters the GAC is unlikely to reduce the disposal issue unless sequestering agents are used to prevent precipitation. Pretreatment, such as with ion-exchange, would just accumulate the long-lived radionuclides on the resin, also presenting a disposal problem.

In addition, Lowry and others (1990), Lowry and Brandow (1985) and Dixon and Lee (1988) have noted that backwashing releases sorbed radon to treated water, although this has not been observed in all cases (Cornwell and others 1999; Kinner and others 1990; 1989). Desorption of the nongaseous radon progeny has not been observed during backwashing (Lowry and others 1990). In this case, disposing of the initial water produced after backwashing and not sending it to the consumer, could eliminate the risk of exposure from release of desorbed radon into the water supply.

Water Storage

Because radon has a relatively short half-life, it is possible to obtain some reduction in concentration by storing the water. It can be stored in separate storage tanks or in those normally used to provide water to a community during periods when demand exceeds the yield of wells. Over 24 h, radon reduction due to storage averages 20-40%, and 5-6 d of storage yields 80-90% losses (Kinner and others 1989). In two waterworks in Sweden, losses of 17-34% were documented during storage (Mjönes 1997). The small reductions mean that this method of treatment is effective only where the percentage removals required to meet the MCL are relatively low and the daily demand for water is small. For example, storage might be an adequate treatment for a school that uses a well to supply water (that is, is not served by a community supply). Repumping is usually required with storage systems used for radon reduction because they are typically operated at atmospheric pressure. Repumping can be avoided if the storage tank is elevated.

Perhaps the biggest problem with this method is providing a reliable and consistent quality of water. As demand fluctuates, the retention time in the storage tank can change, potentially resulting in smaller radon reductions. To avoid that problem, the capacity of storage needs to be increased to ensure acceptable overall radon removal. The tanks usually are vented to the atmosphere, which would increase the risk of air emission as with aeration methods. However, this risk would probably be low because use of storage as a treatment method would be limited to very small water supplies that have relatively low radon concentrations. If radon loss is due solely to decay and not to losses to the atmosphere, there is also the risk of exposure from ingestion of radon progeny such as ^{210}Pb. Because the storage tanks are typically vented to the atmosphere, they might require disinfection under the pending GWDR. As a result, there could be an increased risk of exposure to disinfection byproducts if chlorination is used.

Simple Aeration During Storage

Losses observed after water passes through a storage tank are often higher than those measured when the water resides in the tank undisturbed (Mjönes 1997; Kinner and others 1989). The increased loss has been attributed to aeration that occurs when the water splashes into the tanks. Indeed, the mode of entry is very important. Bottom entry below the water line yields removals similar to loss due to storage alone. Free-fall or entry via spray nozzle or splash box can increase removals to the range of 50-70% (Mose 1993; Kinner and others 1987). Adding a crude coarse bubble-aeration system to the tank can boost removal to the 90%

range (Kinner and others 1987). The issues associated with these systems are the same as those outlined for storage alone (acceptable percentage removals in spite of variable retention times) and for aeration systems (disinfection byproduct formation upon disinfection, off-gas emissions, and precipitate formation).

Blending

Some public supplies (for example, for some large communities) use a combination of surface water and groundwater. In this case, radon-free or very-low-radon water might be available to mix with the groundwater, which would result in reduction in concentration due to dilution. Blending has been documented as effective in some water supplies (Kennedy/Jenks 1991b; Dixon and Lee 1988), but reductions are usually low (20%) (Drago 1998). The use of blending depends on whether the water is monitored for compliance as it enters the distribution system or when it reaches the first tap (consumer). In the former case, both types of water would need to be available at the same location. In addition, blending can be effective as a best available technology only if mixing is complete.

Reverse Osmosis

In the 1991 proposed rule, EPA specified reverse osmosis (RO) as a best available technology for uranium, radium, and beta- and photon-emitters, but not for radon. RO systems use semipermeable membranes and pressure to separate dissolved species from water. In Sweden, Boox (1995) used an RO filter to treat water in two homes. The systems were installed to improve the taste of the water, but they concurrently reduced the radon content by 90%. RO systems have not been tested for radon removal in the United States, and their use in Sweden exclusively for treatment of radon is doubtful because of their low capacity and relatively high cost. In addition, RO membranes do not work well if turbidity-causing material or precipitates (for example, iron, manganese, silica, and calcium) foul them in the raw water. A brine that contains the contaminants removed from the water is created and must be disposed of. Because the brine is concentrated, the levels of radionuclides might be high, dictating special disposal (as outlined for GAC and aeration systems).

Loss in the Water Distribution System

There have been several studies of this method of radon removal from drinking water. The majority of the decrease in radon results from decay during transit or storage in the closed piping network. Most of the studies have documented losses in the range of 10-20% (Rand and others 1991; Kinner and others 1989; Dixon and Lee 1988). A study in Sweden of four waterworks (Mjönes 1997) found higher losses (30-70%), but some of this could have been due to volatilization during pumping or mixing and agitation.

If EPA requires that water meet the MCL when it enters the distribution system, this method of treatment would not be acceptable. Furthermore, it is doubtful that it would consistently produce water with the same radon content, because retention times and therefore losses differ on the basis of the distance of travel and the demand. This method probably would be used only where the radon levels entering the distribution network are relatively low.

Field and others (1998; 1995) have shown that radon levels in a distribution system can actually increase when radon is released during decay of radium associated with iron-based pipe scale. A generation rate would have to be quantified if loss in the distribution system were considered as a treatment alternative in water supplies that have this type of scaling problem. It also has implications for where water samples are collected (at the origin of the distribution system or at the point-of-use) (Field and others 1995).

Vacuum Deaeration and Hollow-Fiber Membrane Systems

These are very new systems that have undergone only laboratory- or pilot-scale testing (Drago 1998; 1997). They have been developed to address the issue of off-gas emissions associated with aeration systems. In both technologies, the radon removed is trapped in a sidestream of water rather than being released to the air. Therefore, these systems have the potential to fill a niche where radon concentrations in the raw water are high (precluding use of GAC) and air emissions of radon are prohibited (precluding use of aeration systems alone). The sidestream water is passed through a GAC bed that sorbs the radon. The GAC is effective in these cases because the radon is dissolved in water, not in a vapor phase. Descriptions of vacuum deaeration and hollow-fiber membrane systems are found in appendix C.

The issues of gamma emissions and disposal of the spent GAC used in the vacuum deaeration and hollow-fiber membrane systems are similar to those for GAC when it is used directly to remove radon from water. Though disinfection might be required to prevent biofilm development on the GAC, microbial and disinfection-byproduct risks are not applicable, because water from the GAC unit used in vacuum deaeration and hollow-fiber membrane systems will not be released to the consumer. The issues of precipitate accumulation and backwashing are also minimized because the sidestream water can be fairly clean. The low transfer efficiency of radon from the sidestream water to the GAC dictates a long EBCT. This, along with the complexity of the systems, would increase the costs of these systems.

CONCLUSIONS

• Several aeration methods exist or are being developed to remove radon from drinking water. Aeration was designated as the best available technology by EPA in its 1991 proposed radon rule; however, some issues and secondary effects are associated with these technologies. They discharge radon into the air (intermedia pollution), and the extent to which the off-gas emissions will be regulated is not clear. Removing radon from the off-gas is much more difficult and expensive than removing it from water. Although aeration is, in general, a straightforward technology to use, aeration of groundwater that contains dissolved species, such as iron, can lead to the formation of precipitates (scale) that can cause operational problems, aesthetic concerns, and disposal problems (for example, iron precipitates enriched with long-lived radionuclides).

• GAC was not listed as a best available technology by EPA in the 1991 proposed radon rule, but it might be an option for small systems that require only minor treatment to meet the MCL. Several issues and secondary effects are associated with GAC. The primary one is retention of radionuclides, which can lead to substantial gamma emissions from the unit and pose a potential problem of radiation exposure for plant operators. Equivalent gamma doses from GAC units that remove radon from public water supplies should be predicted with an extended-source model that can address more-complex geometries. Accumulation of radionuclides can also lead to potential disposal problems for the spent GAC. Both those issues are most problematic when radon loadings are high (that is when treated flows are high or large amounts of radon are retained on the GAC).

• The treated water leaving aeration and GAC water-treatment systems might require disinfection on the basis of the pending GWDR. If chlorination is used as the disinfection method, trihalomethanes, which are known to cause cancer in rats and mice, could be created. The committee estimated that the average lifetime risk associated with disinfection byproducts formed during chlorination of groundwater is on the order of 5×10^{-5}.

• Storage, blending, loss in the water-distribution system, and other newly emerging technologies (such as vacuum deaeration and hollow-fiber membrane systems) are not likely to be major alternatives for removal of radon from drinking water. Storage, blending, and loss in the water distribution system would be limited to situations where only low removal would be required.

9

Multimedia Approach to Risk Reduction

The 1996 Safe Drinking Act Amendments permits states to develop a multimedia approach to reduce the health risk associated with radon if the maximum contaminant level (MCL) were so stringent as to make the contribution of waterborne radon to the indoor radon concentration less than the national average concentration in ambient (outdoor) air. Under those circumstances, an alternative maximum contaminant level (AMCL) would be defined as the radon concentration in water that results in a contribution of waterborne radon to the indoor air concentration equal to the national average ambient radon concentration. The Administrator of the Environmental Protection Agency (EPA) is required to publish guidelines, including criteria, for establishing multimedia approaches to mitigate radon levels in indoor air that will result in an equivalent or greater reduction in the health risk posed by radon in the area served by a public water supply that contains radon in concentrations greater than the MCL but less than or equal to the AMCL.

The objective of this chapter is to describe the considerations involved in performing a quantitative evaluation of the health-risk reductions that would be achieved by reducing the concentration of radon in the water relative to those achieved through such multimedia activities as mitigating homes to reduce their average indoor radon concentrations. To consider how the multimedia approach to risk reduction might be applied, the committee provides several scenarios that suggest how the risks posed by radon in the indoor air of a home from soil gas can be compared with the risks posed by radon in the drinking water of that dwelling.

DERIVATION OF THE ALTERNATIVE MAXIMUM
CONTAMINANT LEVEL (AMCL)

The AMCL for radon is defined as the concentration of radon in water that will contribute to indoor air a radon concentration equal to the national average ambient-air concentration. In effect, the AMCL is defined implicitly by the following relationship:

$$(AMCL) \ (TF) = M_{ambient}, \qquad\qquad 9.1$$

where TF represents the water-to-indoor-air transfer factor, and $M_{ambient}$ is the national average ambient-air concentration. Note that $M_{ambient}$ is a single number that is unknown but that can be estimated with some degree of uncertainty (see chapter 2). The AMCL is also a single number that is to be determined. The role of the transfer factor, TF, in this relationship is less clear, because it is not simply a single number. In fact, the TF is subject to both variability (that is, variation from dwelling to dwelling and over time for a given dwelling) and uncertainty (that is, the distribution of TF over the population of all dwellings is unknown). The legislation that mandates the derivation of an AMCL does not specify how the TF is to be derived. The committee chose to interpret TF as the mean transfer factor (M_{TF}) that is, as a single numerical quantity that, like $M_{ambient}$, is unknown but can be estimated with uncertainty:

$$(AMCL) \ (M_{TF}) = M_{ambient}. \qquad\qquad 9.2$$

The AMCL derived from this relationship has the property that water with radon at the AMCL will, on average over the population of dwellings (that is, over the distribution of TFs), contribute a concentration of $M_{ambient}$ to indoor air. It does not, however, imply that in any given dwelling, the contribution to the airborne radon concentration from water at the AMCL will equal $M_{ambient}$.

The available data regarding ambient-air concentrations and transfer factors are limited but adequate for estimation of M_{TF} and $M_{ambient}$, and the committee determined that the best estimate of the AMCL that can be derived from the available data is the ratio of the estimated arithmetic means:

$$AMCL_{est} = M_{ambient} \ / \ M_{TF} \qquad\qquad 9.3$$

There are, of course, uncertainties in the estimates of both M_{TF} and $M_{ambient}$, so it is necessary to assess the magnitude of uncertainty that they induce in the estimation of the AMCL. If the uncertainties are relatively small, the value of $AMCL_{est}$ is unlikely to differ greatly from the true value defined implicitly by equation 9.2. In view of the limited nature of the data available for estimation of M_{TF} and $M_{ambient}$, the committee chose not to attempt to represent the uncertainties in those estimates with probability distributions. Instead, the committee defined upper and lower bounds that, in its opinion, are highly likely to contain the true values.

As described in chapter 2, the national average ambient air concentration was estimated to be 15 Bq m^{-3} with a high level of certainty that the value lies between 14 and 16 Bq m^{-3}. Therefore, the uncertainty in the estimation of M$_{ambient}$ was represented by lower and upper bounds of 14 and 16 Bq m^{-3}, respectively.

Regarding the mean transfer factor, M$_{TF}$, the committee noted that the compiled measurement data had an estimated mean and standard error of 0.9×10^{-4} and 0.1×10^{-4}, respectively, on the basis of 154 observations, whereas the estimate derived from modeling was either 0.9 or 1.2×10^{-4} (see chapter 3). The committee's best estimate of the mean transfer factor was 1×10^{-4}.

The uncertainty in the estimation of M$_{TF}$ was represented by lower and upper bounds of 0.8×10^{-4} and 1.2×10^{-4}, respectively. Therefore, on the basis of estimates of m$_{ambient}$ (15 Bq m^{-3}) and m$_{TF}$ (1×10^{-4}) given above, the committee estimates the AMCL to be 150,000 Bq m^{-3}. The uncertainty of AMCL$_{est}$ arising from the uncertainties in estimation of M$_{ambient}$ and M$_{TF}$ was estimated by considering the extremes of the bounds that were defined for the two input values. By propagating the upper and lower bounds on the numerator and denominator, the lower bound of the AMCL$_{est}$ is 117,000 Bq m^{-3}, and the upper bound is 200,000 Bq m^{-3}.

EPA will set the MCL value on the basis of the committee's risk assessment in this report and its own policy considerations. However, for the examples in this chapter, it is necessary to assume a value of the MCL. It will be *assumed* that the MCL will be 25,000 Bq m^{-3} of water. The committee makes no recommendation or endorsement of a specific value and is using this assumption only in order to provide a framework for the following discussion of potential risk-reduction scenarios for implementing a multimedia mitigation program.

EQUIVALENT RISK-REDUCTION SCENARIOS

To estimate the health-risk reductions that are obtained by treating water to remove radon or mitigating homes to reduce indoor ^{222}Rn, it is necessary to consider all the associated risks. For example, the processing of water to remove radon would probably then require that the water be disinfected under the proposed rules (EPA 1992b). Disinfection could be performed through illumination with ultraviolet light for small systems or the addition of chlorine or ozone for larger systems. The risks arising from exposure to disinfection byproducts were discussed in chapter 8 and are estimated to be smaller than the risks arising from airborne radon. Thus, the incremental risk posed by disinfection byproducts will not be included in the risk-reduction analysis. The cancer risks to the body associated with ingested radon (2×10^{-9} Bq^{-1} m^3) are small but not negligible when compared with the risk to the lungs posed by the airborne decay products arising from radon released by water used in the home (1.6×10^{-8} Bq^{-1} m^3). Thus, the committee has assumed a mitigation of airborne radon equal to 113% of the airborne radon would provide an equivalent health-risk reduction to account for the risk of radon in the drinking water. This concept will be illustrated in the following scenarios.

Scenario 1: High Radon Concentrations in Water

In scenario 1, the radon concentration in drinking water exceeds the AMCL. The water utility will be required to install treatment equipment to reduce the concentration of radon in the water to at least the AMCL. The incremental cost of further treating the water so that it achieves the MCL will generally be sufficiently small that the multimedia-mitigation approach would probably not be considered. Some additional considerations arise from the increased quantity of radon being removed from the water, such as increased gamma-ray exposure to the water-treatment workers from a GAC bed or the airborne radon released to the atmosphere by an aeration system at the water-treatment plant. In general these factors would not produce sufficient cost differences between meeting the AMCL and meeting the MCL to constitute an incentive to consider a multimedia mitigation program. Thus, the only cases where it is of practical interest to consider implementation of a multimedia program is for water systems in which the radon concentration in the water is between the MCL and the AMCL.

To provide a perspective on how the risk reductions could be compared, we provide an illustrative calculation. Suppose that a water supply contains radon at 125,000 Bq m^{-3}. If the water is treated to reach the assumed MCL, it would provide an average reduction of 125,000 – 25,000 = 100,000 Bq m^{-3}. Multiplying this value by the transfer coefficient of 10^{-4} yields a decrement of 10 Bq m^{-3} in radon concentration in the air in each dwelling. For a water supply that provides water to 1,000 homes each with the same average number of occupants, the committee assumed that there were to be three persons per home. The total reduction in radon resulting from the mitigation of the water to reach the MCL would be 10 Bq m^{-3} per dwelling × 1,000 dwellings, or a cumulative reduction within the community of 10,000 Bq m^{-3}. Taking into account the additional ingestion risk, it would require a reduction of 11,300 Bq m^{-3} in indoor airborne radon to provide the equivalent health-risk reduction.

This risk-reduction analysis could be based on the actual number of occupants in the homes so that the health-risk reductions would be applied to defined populations. However, enumeration of the people in each home presents a potential problem in that the number of individuals in a given dwelling can vary. Homes are sold to new families. Children grow up and move away, and there is the question of the presence or absence of smokers in a home. Because the lung cancer risk posed by radon is significantly higher for smokers than for nonsmokers, a greater health-risk reduction would be obtained by preferentially mitigating the homes of smokers relative to the homes of nonsmokers. Thus, a potential difficulty in demonstrating the continuing benefits of mitigation of homes for health risk reduction is the variability in the number and nature of occupants of the dwelling. The committee has examined health-risk reduction only for an average dwelling within the community.

It is necessary to consider that each home contains the average number of individuals and the same fraction of smokers. If 113 homes were found with high concentrations of airborne radon (concentrations in excess of 150 Bq m^{-3}, the EPA guidance concentration) and these were mitigated so that the average long-term ^{222}Rn reduction in each home were 100 Bq m^{-3}, then the mitigation of these dwellings would provide the same level of risk reduction as reducing the radon in the drinking water to the MCL (100 Bq m^{-3} per dwelling × 113 dwellings = 11,300 Bq m^{-3}). If a typical home mitigation costs \$1,500, the mitigation of the 113 homes costs about \$170,000 plus the estimated cost of home testing (1,000 homes in the community × \$75 per home = \$75,000), for a total of at least \$245,000, which includes the cost of distributing detectors, collecting them, and analyzing the resulting data.

Because of the nature of water-quality regulations, there would be a require-ment for continued monitoring to ensure continuing compliance with the equiva-lent health-risk reduction. Thus, there are O&M costs which would involve an-nual measurements in the mitigated homes and replacement of fans that fail. The typical mean time to failure for the fans is estimated to be about 10 years. Thus, some fans would probably have to be replaced each year. In this scenario, all the homes with concentrations above 150 Bq m^{-3} are to be found and mitigated. To obtain a high level of participation, it would be necessary to attempt to measure the activity concentration in each home with a long-term detector that leads to the high estimated costs to perform the radon survey. The cost to mitigate the water in a community of about 3,000 individuals is estimated by EPA (1991b) to be \$78,000 plus annual O&M costs of \$3,000. Mitigation of radon in indoor air in the 113 homes is substantially more than the cost of buying and operating the system to aerate the water to remove the radon. Thus, the water-supply utility would not choose to adopt the multimedia approach to risk reduction rather than fully mitigate the water to the MCL. However, the American Water Works Asso-ciation estimate (Kennedy/Jenks 1991a) for the acquisition of an appropriate water-treatment system is \$275,000 plus annual O&M costs of \$23,000, so mul-timedia mitigation might be considered as a cost-effective alternative.

Scenarios 2-4: Effects of Distribution of Radon in Indoor Air

On the basis of previously described scenario, an important consideration in deciding on the feasibility of the multimedia approach relative to the water-treatment approaches is whether a subpopulation of dwellings can be identified that would provide the needed equivalent health-risk reduction when their air-borne radon concentrations were reduced sufficiently. Rather than mitigate all the homes that exceed 150 Bq m^{-3}, it would be more cost-effective to mitigate only enough homes to achieve the target level of risk reduction. The prevalence of high-concentration homes will depend on the geology of the area. EPA has separated the United States into three regions of different radon potential (Marcinowski and White 1993). To examine the feasibility of the selective-mitigation approach, we examine the concentration distributions for each of the different radon-potential areas of the United States. In 1989-1990, the EPA conducted the National Residential Radon Survey, NRRS (Marcinowski and others 1994) which provided a statistically valid survey of the distribution of indoor radon concentrations in homes. Each home in the survey was classified by the EPA radon potential region associated with its location. The results are summarized in figure 9.1 for these 3 regions and the entire United States. The lines in the figure were obtained by fitting a lognormal distribution to measured concentrations from each of the three regions.

Scenario 2: Low Radon Potential

To illustrate the problem of finding homes to mitigate, suppose our water supply is in a region of low radon potential. By taking the parameters of the

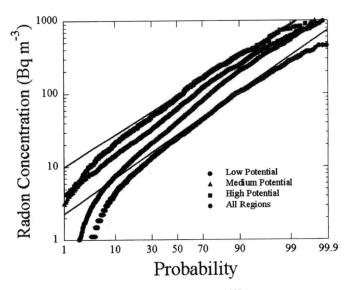

FIGURE 9.1   Parameters of the fitted distributions of ^{222}Rn concentrations for homes in the United States based on the NRRS data (Marcinowski and others 1994). Parameters of distributions are as follows.

	Geometric Mean (Bq m^{-3})	Geometric Standard Deviation
Low Potential	18.2	2.7
Medium Potential	41.8	2.9
High Potential	54.5	2.7
All Data	24.9	3.1

lognormal distribution fitted to the low radon potential region data, five sets of values for the hypothetical 1,000 homes were estimated using a random number generator. Based on the average values of the 5 sets of values, only 2.5% of the homes exceed the 150 Bq m^{-3} indoor-air guidance value suggested by EPA. Thus, only 25 homes would be likely candidates for mitigation of airborne radon. These highest concentration dwellings have an average radon concentration of about 215 Bq m^{-3}. If all were mitigated and the indoor airborne ^{222}Rn concentration were reduced to 75 Bq m^{-3}, a concentration that has been found to be generally achievable with active mitigation systems, it would produce about 3,500 Bq m^{-3} in total radon-concentration reduction.[1] Thus, the multimedia approach could be considered only if the ^{222}Rn concentration in the water were below 60,000 Bq m^{-3} [(60,000 − 25,000)(10^{-4})(1,000) = 3,500], and it would require that all the high-concentration homes could be found and mitigated. For lower concentrations of radon in the water that still exceed the MCL, it would be possible to identify a subset of dwellings that would provide sufficient reduction in airborne radon in enough dwellings to provide an equivalent or greater health-risk reduction. However, it appears likely that the costs of identifying and mitigating enough of dwellings to provide the equivalent health-risk reduction would exceed the costs of processing the water to reduce the radon concentration to the MCL.

Scenarios 3 and 4: Medium and High Radon Potential

For the medium- and high-potential regions, sets of values were generated in a similar manner. For these regions, the fractions of dwellings with more than 150 Bq m^{-3} are about 11% and 20%, respectively, on the basis of the NRRS distributions. The average concentration in the dwellings with concentrations of 150 Bq m^{-3} or more is 250 Bq m^{-3} in the medium-potential region and 270 Bq m^{-3} in the high-potential region. Thus, the problems of finding high-concentration dwellings and reducing their indoor airborne ^{222}Rn concentrations are much smaller than in the low-potential region because there will now be 110 and 200 homes, respectively, that exceed the EPA guidance value and are candidates for mitigation. In the medium-regions, there would be over 19,250 Bq m^{-3} that could be mitigated; in the high-potential area, 39,000 Bq m^{-3} would be available. Thus, the mitigation of a fraction of the homes that exceed the current EPA guidance level would actually produce a larger health risk reduction than mitigation of the water would provide even if the radon concentration in the water supply approached the AMCL.

The ability to obtain the required health-risk reduction by mitigating fewer homes might make the multimedia approach more financially attractive. For the medium-potential region, mitigating the 38 homes with the highest concentra-

---

[1](215 − 75) * 25 = 3,500 Bq m^{-3}.

tions would provide the 11,300 Bq m^{-3}. For the high-potential region, mitigation of the 23 highest-concentration homes would provide the required health-risk reduction. However, in both cases, to obtain the absolute minimum number of dwellings to be mitigated would require that 100% of the dwellings in the community be monitored to ensure finding the highest concentration homes. It is unlikely that such a high level of participation can be achieved, so alternative strategies would need to be adopted. Because there are more high-concentration dwellings to find, an extensive but not exhaustive survey of the community could identify enough high concentration homes to provide the needed health-risk reduction at a cost that would be less than the cost of implementing and maintaining a water-treatment facility.

One cost-effective approach to solicit participation would be to send a notice to ratepayers with their water bills asking whether they know what their indoor radon concentration is, and that if it is above 150 Bq m^{-3} in the home, they might be eligible for mitigation at no cost. The solicitation could also indicate that if the owner were interested in participating, a free test kit would be provided. It is essential that long-term monitoring of radon concentrations be performed in order to provide a reliable estimate of the risk reduction potential. This approach might provide a utility with an initial indication of the availability of high-concentration homes that could be used in developing a health-risk reduction plan. It is the committee's judgment that such an approach is unlikely to identify all the homes that would have to be mitigated to provide an equivalent health-risk reduction, but it would provide a cost-effective way to test the possibility of using the multimedia approach in a utility's operating region.

Scenario 5: Use of New Radon-Resistant Construction

As discussed in chapter 8, the effectiveness of radon-resistant construction is highly uncertain. The committee feels that it is not now possible to quantitatively assign radon-risk reduction potential to such construction practices. In many areas of the country, home construction is not contributing a substantial number of new dwellings to the community. To take credit for using radon-resistant techniques, new houses would have to be connected to existing water supplies. If in the future, the extent of radon reduction in new radon-resistant homes could be reliably estimated, then the following framework could be used to incorporate it into a multimedia mitigation program.

Radon-resistant construction will reduce the indoor radon concentration to a fixed fraction of the value it would have been if conventional construction practices had been used. Thus, it is necessary to estimate what the concentrations would have been in the new homes if they had not been built to be radon-resistant. The potential for radon in these homes will depend on the geology of the area. Assuming that the geology of the area is reasonably uniform, so that existing homes are on geologically comparable soils, a statistically valid survey of long-term average indoor radon concentrations in existing homes would provide the baseline distribution of indoor radon concentrations for the area. Long-term measurements of the radon concentrations in new radon-resistant homes would provide the distribution of radon concentrations in these homes, and the difference between the two distributions would yield the quantitative estimate of the health-risk reduction provided by the construction of new radon-resistant homes. Although the current approaches to radon-resistant building codes are being applied in the EPA high-radon potential areas, only about 20% of the homes in such a region would be expected to exceed the 150 Bq m^{-3} guidance level as previously discussed. Newer methods to estimate the indoor radon potential are likely to provide a basis for refining the regions in which radon-resistant building codes will have the greatest applicability (Price 1997 and references therein). However, it will still be essential to conduct a baseline survey to provide a sound basis for estimating the radon risk reduction potential.

The incremental cost of adding radon-resistant components to new homes is estimated to be $400 per home, so the payment of incentives to new-home contractors to make homes radon-resistant could be economically competitive with water treatment. Thus, the development of a scientifically based estimate of the effectiveness of radon-resistant construction in reducing indoor radon concentrations is critical to being able to use this approach to provide equivalent radon-risk reduction. It is important to note that overall reduction in indoor radon concentrations are not likely to exceed a factor of 2 to 3 (based on the FRI research described earlier), so these techniques may not always result in indoor radon concentrations below the EPA 150 Bq m^{-3} guideline.

Scenario 6: Multicommunity Mitigation

Because the objective of the multimedia-mitigation strategy is to provide equivalent or greater public-health benefit (health-risk reduction) for a lower cost, a scenario could be developed in which a water utility operates wells in several communities or separate production and distribution systems within a single community. Suppose that one community has water with a radon concentration between the MCL and the AMCL and another community has low radon concentrations in its water but high radon concentrations in its indoor air. Could the water utility mitigate the air concentrations in dwellings in the low-water-radon community to produce the equivalent health-risk reduction that would have been obtained by lowering the water concentration of ^{222}Rn in the other community?

The philosophy of maximizing the public-health by using the assumption of linearity in the risks that arise from exposure to radon and its decay products would support this tradeoff as providing cost-effective equivalent risk reduction. On that basis, the committee cannot eliminate this type of risk trading from consideration because it will produce equivalent or better health benefits. However, important questions of equity in the treatment of the two communities must be taken into consideration in the decision as to how to proceed.

A similar scenario can be envisioned in which some homes in a community are served by a public water supply and others have private wells. Under the provisions of the Safe Drinking Water Act (SDWA), the utility would be required only to provide water that meets the radon MCL to the homes that it serves. It is possible that the homes served by private wells would have some of the highest indoor-air radon concentrations. In a holistic view of achieving a comparable or greater health-risk reduction for the community, it might be best to remediate the air in the homes with the highest radon concentrations even if they are not served by the utility. However, that would present a dilemma for the utility because it would be mitigating homes to which it does not provide water. Such dwellings are outside the normal jurisdiction of the SDWA and therefore potentially outside the purview of a multimedia program. A policy decision would be needed as to whether such dwellings could be included in a multimedia mitigation program and would raise an important equity question, in that water ratepayers would be charged for the mitigation of homes that are not being served by their utility and whose occupants are not contributing to the payment of the costs of the radon-abatement program.

Scenario 7: Use of Outreach, Education, and Incentives

Another possible approach to reducing the indoor air concentrations of radon is to enlist homeowners in the identification and mitigation of homes with high radon concentrations. As previously described, home-mitigation programs will be practical only in areas of medium or high indoor-air radon potential or in communities with radon concentrations in the water supply that are close to the MCL. In this case, the utility might involve the community via a public-education program and potentially provide incentives for mitigation of those homes. The committee was asked to comment on the body of evidence regarding the effectiveness of such programs and on how the health-risk reductions could be evaluated in such cases. With respect to outreach and education programs, there is some experience that can be examined.

Communicating risk to the public such that individuals are motivated to change their behaviors and reduce their exposure to the hazard is a well-known problem. The report, *Improving Risk Communication* (National Research Council 1990b), addresses many of the issues relevant to that process. In particular, the report gives an example of comparing radon with other types of risk: "radon risk can equal or exceed the 2% risk of death in an auto accident . . . for anyone who lives 20 years at levels exceeding about 25 picocuries per liter." This statement places an unfamiliar risk (radon exposure in homes) in juxtaposition to a more familiar risk (death in an auto accident). Though such techniques may help people understand the magnitude of an unfamiliar risk, it can also be misleading because it does not specify the respective or acceptable levels of exposure, leaves out potentially relevant nonlethal consequences, and uses language (picocuries per liter) unfamiliar to most people. The comparison above is an example of an expert's message that is precise and accurate but is too complex, or uses unfamiliar technical jargon, such that only another expert would likely understand it. In contrast, simplified messages that nonexperts can understand usually present only selected information, thus, they can be challenged as inaccurate, incomplete, or manipulative.

There have been limited studies of the effectiveness of communicating the risk of radon to the public though little has been peer-reviewed and openly published. A useful summary of state programs to determine the effectiveness of radon programs in mitigating individual risk was prepared by the Conference of Radiation Control Program Directors (1996). Following various types of state outreach programs, CRCPD determined through surveys that a total of 73% of the participants recognized radon, 52% considered radon to be unhealthy, and 44% defined radon correctly. Only 10% of the survey participants tested for radon in their homes, and only 16% of those who thought radon was unhealthy tested their homes. Surveys revealed that radon tests took place for other reasons. For example, 26% of the tests were associated with real-estate transactions though 18% of the tests were carried out despite that residents did not believe radon was unhealthy. The CRCPD surveys indicated that states with radon testing as part of their real-estate transactions requirements also had high-awareness. Furthermore, the surveys indicated that home mitigation was lower for homes with indoor air concentrations less than the EPA-recommended action level of 150 Bq m^{-3}.

It is clear from the CRCPD surveys that certain state radon programs were more effective in communicating risk to the public and that Maine had the greatest success. In Maine, a total of 5% of all homes have been mitigated and 30% of homes with radon levels above average have been mitigated. Of homes that had radon in water at over 370,000 Bq m^{-3}, more than 75% have been mitigated, even though the state of Maine has recommended for the last 20 years that homeowners should mitigate water at levels greater than 740,000 Bq m^{-3}. Factors in Maine which seem to be related to that success include partnering between state agencies and local groups and authorities, and effective use of the media. Wyoming also had high awareness; similar partnering activities were used there as well as using the media to promote radon awareness including outdoor advertising on billboards, direct mail, newspapers, and television publicity. The District of Columbia and Texas had the largest increases in radon awareness at 6% each. The District of Columbia distributed radon information in English and four other languages.

Mitigation of residences for high radon levels varied among the states from 3.6% in Pennsylvania to 0.3% in Hawaii. The states with the highest mitigation rates also were the top 25% of states in terms of public radon awareness. These states provided advice and assistance by telephone as well as printed materials, brochures, and do-it-yourself guides. Three factors were important to influencing the number of households that mitigated radon: educating the public about mitigation and ensuring availability of qualified contractors, a radon-awareness campaign, and promoting the widespread testing of residential radon levels.

Several studies have described the problems of communicating risk to the general public; a broad review of radon-related risk communication was done by the EPA's Science Advisory Board (1995). A telephone survey was used to assess information about homeowners with indoor air concentrations greater than 740 Bq m^{-3} (Field and others 1993). Of these homeowners, only 19% identified lung cancer as a possible health outcome of high radon exposure, and fewer than one-third remembered the value measured in their home to within 370 Bq m^{-3}, even within the first 3 months after receiving their test results. In another study 99 homeowners were randomly selected in a community. In this group, 64% expressed concern about radon but only 7% tested their homes (Kennedy and others 1991). These findings tend to show that knowledge about the hazard does not necessarily lead to actions to reduce the risk. A survey of 275 adults showed that 92% had heard of radon and believed it to be a health risk though only 4% believed that they were exposed to high levels of radon gas (Mainous and Hagen 1993). The phenomenon of believing that exposure happens only to others appears common. In this study, younger and less-educated people were more likely to perceive radon as presenting a health risk and women were found to be 3.5 times as likely as men to perceive radon as a risk. Finally, Sandman (1993) showed that among 3,329 homeowners, the likelihood of radon testing was predicted by the degree of general knowledge about radon and a decision to test was related to each individual's perception of the seriousness of the risk.

Three states have detailed results of testing and mitigation programs. New Jersey examined the short-term home radon-test results, including real-estate tests, by month from 1991 to 1997 (J. Lipoti, State Radiation Protection Program of NJ, private communication). Non-real-estate tests make up about 25% of all the tests for radon in houses. When high radon was found in a test, especially airborne concentrations in excess of 3,700 Bq m^{-3}, free radon packets were sent to homeowners within a 1 mile radius. Roughly one fourth of the homes that received packets used them. These tests were done from January 1996 through April 1997. Another investigation was of the fraction of homes testing high for radon that were not mediated by state-certified contractors. These homes were not reported to have been mitigated. For houses found to have indoor radon concentrations greater than 150 Bq m^{-3}, the percent of dwellings not mitigated ranged from 64 to 72% during the period 1992-1996.

New York state tested radon awareness, testing, and remediation with a survey (NYDH 1997b). The survey included information about ethnic background, age, education, and income and involved more than 1,000 interviews. Of 993 respondents, 152 had tested for radon and 12 had radon concentrations in excess of 150 Bq m^{-3}. Of those in the high-radon group, nine undertook mitigative action and four tested again after mitigation. Those who did not test mainly thought that radon was not high in their home, that radon did not pose a problem in their areas, or that radon risk was exaggerated. When asked to specify their information source, most reported that they had heard about radon on television (27-30%) or from news stories (84-85%). Respondents in high-radon counties showed an increased knowledge about radon over those in low-radon counties by between 50 and 100%. When asked what radon caused, respondents noted headache, asthma, birth defects, lung cancer, and other cancers. Of the 993 respondents who had heard of radon, 86-90% were aware that radon can be unhealthy.

New York state (NYDH 1997a) also performed a mitigation survey. The study sample included 1,522 homes, of which 1,095 had indoor-air radon at over 370 Bq m^{-3} and 427 homes with air concentrations of 150-370 Bq m^{-3}. The subjects in the study were interviewed to ensure that they were at least 18 years old. Of the 1,113 higher-radon-concentration respondents, only 665 (60%) indicated that they had had radon mitigation performed. Mitigation increased with respondent education level from 45% to 65%, increased with household income from 38% to 70%, and increased with household radon level from 47% to 79%. A total of 393, or 59%, of the mitigated homes were retested for radon after mitigation. Respondents where homes were not mitigated had a major concern about the cost of the mitigation. Actions taken included opening windows and doors (51 of 148), sealing or caulking cracks and openings (74 of 148), installing a powered system (9 of 148), installing a system to draw radon (34 of 148), and spending less time in the area with radon (22 of 148). Active mitigation systems were more prevalent for higher radon levels, from 370 to 1,850 Bq m^{-3}. The reasons for performing the mitigation for the people who were most strongly concerned for their own health, concerned for children's health, and the publicity on health effects. Reasons for not performing mitigation were mainly that radon concentration was not too high and that mitigation was too expensive.

Another question is the potential effectiveness of incentives. The costs of a multimedia program could be reduced if homeowners and the utility shared the cost of mitigation. Sweden has a program of partial incentives. It has 8.8 million inhabitants and about 4.1 million dwellings (1.9 million detached houses and 2.2 million multifamily houses). The average radon concentration in Swedish dwellings is 108 Bq m^{-3}. Sweden has a legally enforceable limit for radon in existing dwellings of 400 Bq m^{-3} and a recommendation to reduce radon concentrations that are above 200 Bq m^{-3}. The new-building limit is 200 Bq m^{-3}. Indoor radon concentrations have been measured in about 350,000 dwellings with the cost of the measurement usually paid by the homeowner. About 45,000 homes with radon concentrations above 400 Bq m^{-3} have been found. Based on these measurements, it is projected that about 150,000 dwellings, most of them detached houses, have radon over 400 Bq m^{-3} and about 500,000 have radon over 200 Bq m^{-3}. Of the measured homes with high concentrations (>400 Bq m^{-3}), some 20,000-25,000 have been mitigated. Homeowners can receive a grant for half the cost of remedial work up to 15,000 Swedish krone (SEK) (about $2,000). The homeowners must be able to show that the initial radon concentration exceeds 400 Bq m^{-3}; the measurements must be made in accordance with the protocol for radon measurements in dwellings (issued by the Swedish government), and the intended remedial measures must be approved by the local authorities (288 municipalities). The application is handled by the regional authorities (25 regions). The board for housing and planning has the overall responsibility for the grants. A homeowner who is not satisfied with the decision of the regional authority can complain to the board. During the last 2 y, about 1,900 grants have been paid, at a cost of about 20,000,000 SEK (2.7 million dollars). The Swedish authorities have had advertising campaigns to raise public interest in the radon issue and special campaigns for the grants.

Pennsylvania had a low-interest loan program to encourage homeowners to mitigate their dwellings once they determined that their homes had high concentrations of radon. However, very few of the eligible homeowners took advantage of the program.

The programs described above indicate that public apathy about the potential risks of exposure to radon has generally remained despite numerous and sometimes costly public education efforts. On the basis of these reported results, the committee concludes that an education and outreach program would be insufficient to provide a scientifically sound basis for claiming equivalent health-risk reductions and that an active program of mitigation of homes would be needed to demonstrate health-risk reduction. Nevertheless, education or other programs to deliver basic information about radon could be a useful part of a program to attract homeowners as eventual participants in a mitigation program. Incentives could be used to further increase participation, however, there does not appear to be clear quantitative evidence of the effectiveness of such programs.

Scenario 8: Outreach for Other Health Risks

It has been suggested that because the only effect of indoor radon is lung cancer and the primary cause of lung cancer is smoking, equivalent health-risk reductions could be obtained by an education and outreach program that persuades people to quit smoking. Irrespective of the question of the effectiveness of outreach and education programs, this substitution of causality is a policy issue that is beyond the scope of the committee's charge and expertise.

EQUITY AND IMPLEMENTATION ISSUES AND RISK REDUCTION

Implementation of a multimedia mitigation program presents several potentially major problems for a utility. There are important equity issues that the committee sees as the most critical. Equity issues exist in trading the risks of the entire community against the risks to the occupants of the houses being mitigated. The benefits of the risk-reduction program will go only to the people in the homes that are mitigated rather than to all those who use the water supply and are exposed to radon in the drinking water, so there are questions of fairness that will need to be addressed by the state that establishes the multimedia mitigation program and the utility that implements it. It can be shown statistically that there would be a net public-health benefit to the community if the highest-concentration homes, particularly in the medium- and high-potential areas of the country were mitigated. However, it might be difficult to convince residents whose homes are not treated that the net health benefits to the community, the net economic benefits to the utility, and the benefits to the water-users justify their small increased risk associated with the radon in the water.

Another problem is related to the accounting of the health-risk reduction and the potential natural variability of the indoor concentration of radon. Few homes have been continuously monitored over long periods, but where they have been (Steck 1992), a substantial variability can be observed even in the absence of any changes in construction or in the normal mode of living in the homes. That variability means that there could be increases or decreases in the health-risk reduction obtained by the mitigation of any specific dwelling. It is difficult to assess how much such variability would affect the total aggregate indoor-radon reduction obtained by the mitigation of a number of dwellings. Thus, the committee recommends that a margin of safety be designed into any multimedia mitigation plan. The committee suggests that there be a 10-20% excess in the cumulative amount of indoor radon mitigation performed to ensure that there will always be an equivalent or higher health-risk reduction.

The committee has presented a scenario in which the risks in one community have been traded for the risks in another with a resulting identical or improved public-health effect and a commensurate economic benefit to both communities. Thus, from the viewpoint of public health, it would be reasonable to take the cost-effective solution. However, residents in the community whose water is going untreated, in exchange for reduced risks to those living in what were high-air-borne-concentration homes in the other community, are not likely to be in favor of such a solution even if it does result in a smaller increase in their water costs than would occur if their water were treated. *Thus, non-economic considerations such as equity, fraction of homes mitigated, and other related matters are expected to play a large role in the evaluation of multimedia mitigation programs and might ultimately constitute the deciding factor in whether such a program is undertaken.* In any planning process, a carefully designed program of public education will be essential to provide a perspective on the tradeoffs in the risks being proposed and the health and economic costs and benefits that will be produced by the various alternatives. Because of the sensitivity of the equity issue, the assistance of risk communication experts will be needed in both the planning and implementation stages of public education programs.

Water utilities have traditionally been involved in treating groundwater at the wellhead, or just before its entry into the distribution system. Rarely is water treated by a utility at the tap of the individual home or business because SDWA requirements dictate that water quality be acceptable when water leaves the treatment plant and enters the distribution system, as well as when it arrives at the consumer's tap. Where a decrease in water quality is expected (for example, because of microbial regrowth in the distribution system), a remedy is used to maintain standards (for example, a disinfectant is introduced to prevent regrowth).

If a multimedia approach to the radon problem involves mitigation of air in specifically targeted homes, water utilities will have to oversee the installation, operation, and maintenance of mitigation systems in individual homes. Utilities might have some experience installing, operating and maintaining point-of-entry systems for water in homes, but they are unlikely to have any experience with air mitigation. It is not clear how a water utility, especially a small one, will address this demand for expertise in air mitigation. Many small utilities would have to contract out the installation of the system and then determine how they will monitor the continuing performance at every home. The installation, operation, and maintenance of the airborne-radon reduction systems in individual homes and businesses presents a substantial problem in routinely gaining access to the areas where the treatment units are so that they can be monitored and maintained as required.

Historically, within EPA and many state governments, the personnel addressing issues of airborne and waterborne radon are in different departments, divisions, or even agencies. This division of responsibilities has hindered coordination of policy and response to radon-related issues. The problem is compounded by the fact that waterborne-radon concentrations will be regulated, whereas airborne-radon concentrations are not (only guidelines are provided for indoor-air concentrations). If multimedia approaches to radon are implemented, there will be a need for interaction between the government entities charged with the regulation of radon in water and those familiar with airborne-radon mitigation. It is clear that multimedia approaches to radon mitigation will be varied, and this will require substantial cooperation within and among EPA, the state agencies involved in airborne- and waterborne-radon mitigation and monitoring, water utilities, and local governments. Thus, major problems in policy implementation will need to be addressed.

Another potential problem can be illustrated by a related example. The Water Pollution Control Act Amendments of 1972 mandated that all communities, at a minimum, achieve secondary treatment of their wastewater. In 1977, Congress modified that requirement to allow communities discharging into marine waters to apply for a waiver of secondary treatment if they could demonstrate that it would cause no adverse effect on the environment. The waiver was intended, in part, to relieve rural villages with very small wastewater discharges in such places as Alaska, of the burden of building and operating secondary treatment facilities. When the first round of waiver applications were submitted, most were from large cities that were well organized and could afford to prepare the required environmental-impact assessment. A similar phenomenon could happen with the proposed multimedia approach to radon mitigation, especially if a state chose not to submit such a program to EPA. In this case, the law allows individual public water supplies within a state to submit their own multimedia mitigation programs. It is likely that many very small water utilities whose water contained radon concentrations exceeding the MCL but not the AMCL could not muster the resources or mount the effort required to propose such a program formally. EPA, state agencies, and perhaps the water associations should develop mechanisms to assist small public water supplies in decision-making regarding multimedia mitigation programs.

In the 1991 proposed rule for radon in drinking water, EPA outlined a set of monitoring requirements for establishing and maintaining compliance with the MCL. The agency recommended that systems that must treat their water for radon be required to sample annually to demonstrate compliance. If the water did not meet the MCL, the sampling frequency would be increased to quarterly until the average of four consecutive samples was less than the MCL. The goal of compliance monitoring is to ensure that there is a continued measurable health-risk reduction due to the removal of radon from drinking water. If a multimedia approach were used in which the air in specifically targeted homes is mitigated for radon, the water utility would have to monitor the indoor airborne-radon concentrations in the mitigated dwellings to ensure a continued measurable health-risk reduction. The monitoring requirement should be similar for any new houses built to be radon-resistant. The committee recommends that air-compliance monitoring be required in each home whose air is mitigated and that these compliance requirements be equivalent to the requirements established in the final rule that regulates radon in drinking water.

SUMMARY

A number of important issues will need to be considered by any state agency or local water utility before it proposes the implementation of a multimedia mitigation program. The ease with which dwellings with high indoor radon concentrations can be found within the utility's service area is important because it will mean that houses that can be potentially mitigated can be more easily identified. In addition there will be a large health-risk reduction associated with each mitigated house. At the same time, the smaller number of houses that are mitigated to obtain the same or greater health-risk reduction as would occur from treating the water will also increase the equity issues in that fewer individuals will benefit from the multimedia mitigation program relative to the number being asked to share the remaining risk. Public education will certainly be needed to obtain a community's commitment to the multimedia program and here again, experts in risk communication must be an integral part of the planning and implementation team. The committee believes that the equity questions that are generated by the program of risk-trading could represent the most important barrier to the implementation of a cost-effective program that yields maximum public-health benefits.

EPA and the state agencies responsible for water quality will continue to be faced with the problem that the health risks arising from the presence of radon in drinking water are essentially associated with the water's contribution to the indoor air concentration. With an average transfer coefficient of 10^{-4}, the increment of indoor radon that emanates from water will generally be smaller than the average concentration of radon already present in the dwellings from other sources. Thus, even if water treatment is required, the reduction of radon in water will not substantially reduce the total radon-related health risks that are faced by the occupants of the dwellings being served by the water utility.

10

Research Recommendations

The committee has identified two general classes of research recommendations that need to be addressed to provide a better basis for radon-risk reduction in the future. The first group contains research issues that are uniquely or very strongly related to the uncertainties associated with radon in drinking water. The second set are broader in scope, more generic, and thus less directly linked to the risks posed by radon in water; however, these other issues are important because they affect the evaluation of radon risks.

The first research recommendation is to provide a better basis for evaluating the risks associated with ingestion of radon in drinking water. In the committee's modeling effort, the risk posed by ingestion of radon depended in large part on the estimation of the extent to which radon diffuses through the stomach wall; no data are available to address this. Although bounds on the risks can be estimated by considering zero or 100% diffusion, the resulting risk estimates extend over a factor of 100. Experiments that limit the range of diffusion would be valuable in providing narrower bounds on the estimated risks. It is also difficult to provide quantitative estimates of the uncertainties in the overall risk. Further information that would permit a more complete quantitative evaluation of the uncertainties in the ingestion risk would be useful in comparing the ingestion and inhalation risks.

The second principal recommendation is to investigate the efficacy of radon-resistant new construction. This is important in the context of the multimedia risk-reduction approach and in the general issue of improving the quality of the science in support of designing and building radon-resistant new buildings. The issues related to estimating the effectiveness of radon-resistant new construction were described in chapter 8 but several research areas which should be considered are noted here. (a) Long-term radon measurements are needed (after occupancy) in a number of radon-resistant houses in several regions for comparison with measurements on "control" houses in the same regions. (b) Success in building effective radon-resistant homes should be correlated with the attention paid to the details of the construction process, as the Florida experience appears to indicate. (c) Soil-gas concentrations should be measured both as a criterion for examining radon resistance and as a means of determining whether there are practical upper limits to the use of radon resistance in limiting indoor concentrations to about 150 Bq m^{-3}.

More specific to the multimedia programs, carefully designed studies are needed on a sufficient number of new homes to provide an adequate basis for quantitatively estimating the health-risk reductions that can be achieved by radon-resistant construction. These studies should be long enough to determine the efficacy and durability of the risk reductions achieved by the package of design features included in a radon-resistant house.

There are several broader research subjects whose study the committee believes would shed additional light on the subjects covered in this report, although they are not likely to alter substantially the overall conclusions presented here. For example, a number of important basic scientific questions regarding the nature of radiation-induced cancers are still unanswered. The most important of these issues is the molecular analysis of the effects of single alpha-particle tracks, including single-cell analysis of DNA damage, DNA repair mechanisms, and the linearity of the dose-response curve under low-dose conditions.

Included in these other research issues are those related to exposure and risk reduction. There is a need for better, more nationally representative data on the water-to-air transfer coefficient, on specific water-use rates in homes, and on home ventilation rates. The interplay between public-health risk perception and the alternative risk-reduction strategies that might be used in a multimedia risk-reduction approach should be better understood.

Furthermore, it would be useful to have better data on the long-term annual average indoor radon concentrations at state or regional levels. A comprehensive, geographically based ambient-radon study that would better incorporate the major populations of the United States and their geologic variability as well as focused regional studies of ambient radon in high radon areas of the country would also be useful. Data from all of these studies would better support the estimation of the national baseline exposure and risk.

Finally, better data on particle size distributions and the resulting indoor exposures is needed. The size of the particles in the indoor aerosol is a key determinant of the deposition of radon decay products in the lungs and of other health risks associated with indoor exposure to particulate matter. Thus, a national, statistically valid assessment of the distribution of human exposures to indoor aerosols would be helpful in many risk-assessment problems, including the effects of the presence of radon decay products in indoor air.

References

Ager D, Phillips JW, Abella Columna E, Winegar RA, Morgan WF. 1991. Analysis of restriction enzyme-induced double-strand breaks in Chinese hamster ovary cells by pulsed-field gel electrophoresis: Implications for chromosome damage. Radiation Research 128:150-156.

Alvarez L, Evans JW, Wilks R, Lucas JN, Brown JM, Giaccia AJ. 1993. Chromosomal radiosensitivity at intrachromosomal telomeric sites. Genes, Chromosomes and Cancer 8:8-14.

Anderson CW. 1993. DNA damage and the DNA-activated protein kinase. Trends In Biomedical Sciences 18:433-437.

Andrews LC. 1986. Elementary Partial Differential Equations with Boundary Value Problems. Orlando, FL: Academic Press Inc.

Arnheim N, Shibata D. 1997. DNA mismatch repair in mammals: role in disease and meiosis. Current Opinion in Genetics and Development 7:364-370.

Ashley T, Ward DC. 1993. A "hot spot" of recombination coincides with an interstitial telomeric sequence in the American hamster. Cytogenetics and Cell Genetics 62:169-172.

AWWA. (American Water Works Association). 1991. Existing VOC Treatment Installations: Design, Operating and Cost, Report of the Organic Contaminants Control Committee. Denver, CO: Water Quality Division.

Bakulin VN, Senko EE, Starikov B. 1970. Investigation of turbulent exchange and wash-out by measurement of natural radioactivity in surface air. Journal of Geophysical Research 75:3669-3674.

Becker III AP, Lachajczyk TM. 1984. Evaluation of Waterborne Radon Impact on Indoor Air Quality and Assessment of Control Options. EPA-600/7-84-093. Springfield, VA: National Technical Information Service: EPA Industrial Environmental Research Laboratory.

Bell CM, Leach MO. 1982. A compartment model for investigating the influence of physiological factors on the rate of washout of Xe-133 and Ar-37 from the body. Physics in Medicine and Biology 27(9):1105-1117.

Bendixsen CL, Buckham JA. 1973. General survey technique for separation and containment of noble gases from nuclear facilities. In: Proceedings of the Noble Gas Symposium. EPA-600/6-76-026. Las Vegas, NV: Environmental Protection Agency, National Environmental Research Center and the University of Nevada. pp. 290-295.

Benjamin MB, Little JB. 1992. X-rays induce interallelic homologous recombination at the human thymidine kinase gene. Molecular and Cellular Biology 12:2730-2738.

Bernard SR, Snyder WS. 1975. Metabolic models for estimation of internal radiation exposure received by human subjects from inhalation of noble gasses. ORNL-5046. Oak Ridge, TN: Oak Ridge National Laboratory.

Bernhardt GP, Hess CT. 1996. Acute exposure from Rn-222 and aerosols in drinking water. Environment International 22(Suppl. 1):S753-S759.

Bettega D, Calzolari P, Chiorda GN, Tallone-Lombardi L. 1992. Transformation of C3H 10T1/2 cells with 4.3 MeV alpha particles at low doses: effects of single and fractionated doses. Radiation Research 131:66-71.

Bischoff KB. 1986. Physiological pharmacokinetics. Bulletin of Mathematical Biophysics 48(3-4):309-322.

Bissonnette RP, Echeverri F, Mahboubi A, Green DR. 1992. Apoptotic cell death induced by c-myc is inhibited by bcl-2. Nature 359:552-554.

Blanchard RL. 1989. Ambient Outdoor Radon Concentrations in the United States. Contract No. 68-02-4375 for USEPA, Office of Radiation Programs. McLean, VA: S. Cohen and Associates.

Bocanegra R, Hopke PK. 1987. The feasibility of using activated charcoal to control indoor radon. In: Hopke PK (eds.), Radon and Its Decay Products, Occurrence, Properties, and Health Effects. Symposium Series 331. Washington, DC: American Chemical Society. pp. 560-569.

Bocanegra R, Hopke PK. 1989. Theoretical evaluation of indoor radon control using a carbon adsorption system. Journal of the Air Pollution Control Association 39:305-309.

Boehm T, Folkman J, Browder T, O'Reilly MS. 1997. Antiangiogenic therapy of experimental cancer does not induce acquired drug resistance. Nature 390:404-407.

Bogen KT, Spear RC. 1987. Integrating uncertainty and interindividual variability in environmental risk assessments. Risk Analysis 7:427-436.

Bond JH, Levitt DG, Levitt MD. 1977. Quantitation of countercurrent exchange during passive absorption from the dog small intestine. Journal of Clinical Investigation 59:308-318.

Boox C. 1995. The Effect of Water Treatment Equipment on the Radon Concentrations in Water From Drilled Wells (in Swedish, English summary). Report 95-14. Stockholm, Sweden: Statens Strålskyddinstitut.

Borak TB, Baynes SA. 1999. Continuous measurements of outdoor ^{222}Rn concentrations for three years at one location in Colorado. Health Physics 76(4):418-420.

Borak TB Johnson JA. 1988. Estimating the Risk of Lung Cancer from Inhalation of Radon Daughters Indoors: Review and Evaluation. EPA 600/6-88/008. Las Vegas, NV: US Environmental Protection Agency.

Bouffler S, Silver A, Papworth D, Coates J, Cox R. 1993. Murine radiation myeloid leukaemogenesis: relationship between interstitial telomere-like sequences and chromosome 2 fragile sites. Genes, Chromosomes and Cancer 6:98-106.

Bowen PT, Harp JF, Baxter JW, Shull RD. 1993. Residential Water Use Patterns. Denver, CO: American Water Works Association Research Foundation.

Brash DE, Rudolph JA, Simon JA, Lin A, McKenna GJ, Baden HP, Halperin AJ, Ponten J. 1991. A role for sunlight in skin cancer: UV-induced p53 mutations in squamous cell carcinoma. Proceedings of the National Academy of Sciences 88:10124-10128.

Brass HH, Weisner MJ, Kingsley BA. 1981. Community water supply survey; sampling analysis for purgeable organics and total organic carbon (draft). In: American Water Works Association Annual Meeting. Water Quality Division.

Brennan T, Osbourne MC, Brodhead B. 1990. Evaluation of radon resistant new construction techniques. In: 1990 International Symposium on Radon and Radon Reduction Technology. Vol. V. Research Triangle Park, NC: U.S. Environmental Protection Agency, Office of Research and Development. Paper VIII-1.

Brenner DJ, Ward JF. 1992. Constraints on energy deposition and target size of multiply-damaged sites associated with DNA double-strand breaks. International Journal of Radiation Biology 61:737-748.

Brenner DJ. 1992. Radon: current challenges in cellular radiobiology. International Journal of Radiation Biology 61:3-13.

Brenner DJ, Sachs RK. 1994. Chromosomal "fingerprints" of prior exposure to densely ionizing radiation. Radiation Research 140:134-142.

Brenner DJ. 1994. The significance of dose rate in assessing the hazards of domestic radon exposure. Health Physics 67:76-79.

Brenner DJ, Miller RC, Huang Y, Hall EJ. 1995. The biological effectiveness of radon-progeny alpha particles. III. Quality factors. Radiation Research 142:61-69.

Brisk MA, Turk A. 1984. Control of Indoor Radon by Activated Carbon Filters, Paper 84-35.3, 77th Annual Meeting of the Air Pollution Control Association. Pittsburgh, PA: Air Pollution Control Association.

Brodhead Bill. 1995. Nationwide survey of RCP listed mitigation contractors . In: 1995 International Radon Symposium. Nashville, TN: American Association of Radon Scientists and Technologists. pp. III-5.1 to III-5.14.

Brown WL, Hess CT. 1992. Measurement of the biotransfer and time constant of radon from ingested water by human breath analysis. Health Physics 62:162-170.

Bruno R. 1983. Sources of indoor radon in houses: a review. Journal of the Air Pollution Control Association 33:105-109.

Buchhop S, Gibson MK, Wang XW, Wagner P, Sturzbechter HW, Harris CW. 1997. Interaction of p53 with the human rad51 protein. Nucleic Acids Research 25:3868-3874.

Burkhart JF, Kladder DL. 1991. Comparison of indoor radon concentrations between preconstruction and postconstruction mitigated single family dwellings. In: The International 1991 Symposium on Radon and Radon Reduction Technology. Vol. IV. Research Triangle Park, NC: US Environmental Protection Agency, Office of Research and Development. Paper VIII-2.

Busigin A, VanDerVooren AW, Phillips CR. 1979. Interpretation of the response of continuous radon monitors to transient radon concentrations. Health Physics 37:659-667.

Caelles C, Helmberg A, Karin M. 1994. p53-dependent apoptosis in the absence of transcriptional activation of p53 target genes. Nature 370:220-223.

Caldecott KW, McKeown CK, Tucker JD, Ljungquist S, Thompson LH. 1994. An interaction between mammalian DNA repair protein XRCCI and DNA ligase III. Molecular and Cellular Biology 14:68-76.

Camper AK, LeChevallier MW, Broadaway SC, McFeters GA. 1985. Growth and persistence of pathogens on granular activated carbon filters. Applied and Environmental Microbiology 50:1378-1382.

Camper AK, LeChevallier MW, Broadaway SC, McFeters GA. 1986. Bacteria associated with granular activated carbon particles in drinking water. Applied and Environmental Microbiology 52:434-438.

Camper AK, Broadaway SC, LeChevallier MW, McFeters GA. 1987. Operational variables and the release of colonized granular activated carbon particles in drinking water. Journal of the American Water Works Association 79:70-74.

Canman CE, Chen CY, Lee MH, Kastan MB. 1994. DNA damage responses: p53 induction, cell cycle perturbations, and apoptosis. Cold Spring Harbor Symposia on Quantitative Biology 59:277-286.

Carslaw HS, Jaeger JC. 1959. Conduction of Heat in Solids. 2nd Ed. Oxford, UK: Oxford University Press.

Castrèn O. 1980. The contribution of bored wells to respiratory radon daughter exposures in Finland. In: Gesell TF, Lowder WM (eds.), Natural Radiation Environment III. Report No. CONF-780422. Springfield, VA: National Technical Information Service. pp. 1364-1370.

Cavallo A, Gadsby K, Reddy TA. 1996. Comparison of natural and forced ventilation for radon mitigation in houses. Environment International 22(Suppl. 1):S1073-S1078.

Chambers DB, Reilly PM, Lowe LM, Stager RH, DuPont P. 1990. Effects of exposure uncertainty on estimation of radon risks. In: Cross FT (eds.), Indoor Radon and Lung Cancer: Reality or Myth? 29th Hanford Symposium on Health and the Environment. Columbus, OH: Battelle Press.

Chen M, Quinats J, Fuks Z, Thompson C, Kufe DW, Weichselbaum RR. 1995. Suppression of Bcl-2 messenger RNA production may mediate apoptosis after ionizing radiation, tumor necrosis factor a, and ceramide. Cancer Res. 55:991-994.

Cheong N, Wang X, Wang Y, Iliakas G. 1994. Loss of S-phase-dependent radioresistance in irs-1 cells exposed to X-rays. Mutation Research 314:77-85.

Chittaporn P, Eisenbud M, Harley NH. 1981. A continuous monitor for the measurements of environmental radon. Health Physics 41:405-410.

Chittaporn P, Harley NH. 1994. Water use contribution to indoor radon (abstract). Health Physics 66(6):S28-S29.

Cleaver JE, Morgan WF. 1991. Poly(ADP-ribose) polymerase: a perplexing participant in cellular responses to DNA breakage. Mutation Research 328:1-16.

Clements WE, Wilkening MH. 1974. Atmospheric pressure effects on Rn-222 transport across the air-earth interface. Journal of Geophysical Research 79:5025.

Clifford DA. 1990. Removal of radium from drinking water. In: Cothern CR, Rebers PA (eds.), Radon, Radium and Uranium in Drinking Water. Chelsea, MI: Lewis Publishers. pp. 225-247.

Cohen BL. 1989. Measured radon levels in U.S. homes. In: Proceedings of the Twenty-Fourth Annual Meeting of the National Council on Radiation Protection and Measurements. NCRP Proceedings No. 10. Bethesda, MD: National Council on Radiation Protection and Measurements. pp. 170-181.

Cohen BL. 1990. A test of the linear no-theshold theory of radiation carcinogenesis. Environmental Research 53:193-220.

Cohen BL. 1992. Compilation and integration of studies of radon levels in US homes by states and counties. Critical Reviews in Environmental Control 22:243-364.

Cohen BL. 1995. Test of the linear no-threshold theory of radiation carcinogenesis for inhaled radon decay products. Health Physics 68:157-174.

Cohen GM. 1997. Caspases: the executioners of apoptosis. Biochemical Journal 328:1-16.

Cornwell DA, Kinner NE, McTigue NE. 1999. Assessment of GAC Adsorption for Radon Removal. Denver, CO: American Water Works Association Research Foundation.

Correia J, Weise S, Callahan R, Strauss H. 1987. The Kinetics of Ingested Rn-222 in Humans Determined From Measurements With Xe-133. Cooperative Agreement CR810942. Research Triangle Park, NC: US Environmental Protection Agency, Health Effects Laboratory.

Correia JA, Weise S, Callahan RJ, Dragotakes S, Strauss W. 1988. Cumulative organ radioactivity concentrations of 222-radon and its progeny following ingestion. Journal of Nuclear Medicine 29(5):872-873.

Couch FJ, Weber BL. 1998. Breast cancer. In: Vogelstein B, Kinzler KW (eds.), The Genetic Basis of Human Cancer. New York, NY: McGraw-Hill.

Cox R. 1994. Molecular mechanisms of radiation oncogenesis. International Journal of Radiation Biology 65:57-64.

Crawford-Brown DJ. 1991. Cancer fatalities from waterborne Rn-222. Risk Analysis 11:135-143.

Crawford-Brown DJ. 1989. The biokinetics and dosimetry of radon-222 in the human body following ingestion of ground water. Environmental Geochemistry and Health 11:10-17.

CRC. 1996. Handbook of Chemistry and Physics. Boca Raton, FL: CRC Press.

CRCPD. (Conference of Radiation Control Program Directors, Inc.). 1994. CRCPD Radon Risk Communication and Results Study. Frankfort, KY: Conference of Radiation Control Program Directors, Inc.

CRCPD. (Conference of Radiation Control Program Directors Inc.). 1996. State Radon Activities: Key Elements, Measure of Success, and Pitfalls, A Report to the CRCPD Committee on Radon (E-25). Frankfort, KY: Conference of Radiation Control Program Directors, Inc.

Cristy M, Eckerman KF. 1987. Specific Absorbed Fractions of Energy at Various Ages from Internal Photon Sources. ORNL/TM-8381/V1-7. Oak Ridge, TN: Oak Ridge National Laboratory.

Cristy M, Eckerman KF. 1993. SEECAL: Program to Calculate Age-Dependent Specific Effective Energies. ORNL/TM-12351. Oak Ridge, TN: Oak Ridge National Laboratory.

Culot MV, Olson HG, Schiager KJ. 1978. Field applications of a radon barrier to reduce indoor airborne radon progeny. Health Physics 34:498-501.

Cummins MD. 1987. Removal of Radon from Contaminated Groundwater by Packed-Column Air Stripping. USEPA Draft Report. Cincinnati, OH: Environmental Protection Agency, Office of Drinking Water, Technical Support Division.

Cunningham RP. 1997. DNA repair: how yeast repairs radical damage. Current Biology 6:1230-1233.

Darby SC, Radford EP, Whitley E. 1995. Radon exposure and cancers other than lung cancer in Swedish iron miners. Environmental Health Perspectives 103(2):45-47.

Datye VK, Hopke PK, Fitzgerald B, Raunemaa T. 1997. Dynamic Model for Assessing 222-Rn and Progeny Exposure from Showering with Radon Bearing Water. Environmental Science and Technology 31:1589-1596.

Day JP, Marder BA, Morgan WF. 1993. Telomeres and their possible role in chromosome stabilization. Environmental and Molecular Mutagenesis 22:245-249.

Deb AK. 1992. Contribution of Waterborne Radon to Home Air Quality. Denver, CO: American Water Works Association Research Foundation.

Dehmel JCF, Brodhead W, Kladder D, Hall S, Mardis M. 1993. Private Mitigation System Durability Study. Contract 68-D0-0097 with USEPA. McLean, VA: S. Cohen & Associates.

Dendy PP, Smith CL, Aebi AE. 1967. The role of peroxides in the inhibition of DNA synthesis following irradiation with a UV microbeam. Photochemistry and Photobiology 6:461-467.

Dewey R, Nowak M, Murane D. 1994. Radon mitigation effectiveness in new home construction: Passive and Active Technologies. In: 1994 International Radon Symposium. Atlantic City, NJ: American Association of Radon Scientists and Technologists. pp. V-1.1 to V-1.8.

Dixon KL, Lee RG. 1988. Occurrence of radon in well supplies. Journal of the American Water Works Association 80:65-70.

DOE. (Department of Energy). 1995. Housing Characteristics 1993. DOE/EIA-0314(93). Washington, DC: US Department of Energy, Energy Information Administration.

DOE/CEC. (Department of Energy and Commission of European Communities). 1989. International Workshop on Residential Radon Epidemiology. CONF-8907178. Washington, DC: Deparment of Energy.

Dogliotti E. 1996. Mutational spectra: from model systems to cancer-related genes. Carcinogenesis 17:2113-2118.

Drago JA. 1997. Radon removal using microporous hollow fiber membranes. In: Proceedings of American Water Works Association Membrane Technology Conference. Denver, CO: American Water Works Association. pp. 831-841.

Drago JA. 1998. Critical Assessment of Radon Removal Systems for Drinking Water Supplies. Denver, CO : American Water Works Association Research Foundation.

Driggers WJ, Grishko VI, LeDoux SP, Wilson GL. 1996. Defective repair of oxidative damage in the mitochrondrial DNA of a xeroderma pigmentosum group A cell line. Cancer Research 56:1262-1266.

Druilhet AD, Guedalia D, Fontan I. 1980. Use of natural radioactive tracers for the determination of vertical exchanges in the planetary boundary layer. In: Gesell TF, Lowder WM (eds.), Natural Radiation Environment III. Report No. CONF-780422. Springfield, VA: National Technical Information Service. pp. 226-241.

Dua SK, Hopke PK. 1996. Hygroscopicity of indoor aerosols and their influence on the deposition of inhaled radon decay products. Environment International 22:S941-S947.

Duval JS, Jones WJ, Riggle FR, Pitkin JA. 1989. Equivalent Uranium Map of the Conterminous United States. Open File Report 89-478. Reston, VA: US Geological Survey.

Dyksen JE, Raczko RF, Cline GC. 1995. Operating experiences at VOC treatment facilities. In: Proceedings of the 1995 Annual American Water Works Association Conference, Anaheim, CA. Denver, CO: American Water Works Association. pp. 659-684.

Eckerman KF, Leggett RW, Nelson CB, Puskin JS, Richardson ACB. 1998. Health Risks from Low-Level Environmental Exposure to Radionuclides, Federal Guidance Report No. 13, Part 1-Interim Version. EPA 402-R-97-014. Oak Ridge National Laboratory and Environmental Protection Agency.

Eguchi Y, Shimizu S, Tsujimoto Y. 1997. Intracellular ATP levels determine cell death fate by apoptosis or necrosis. Cancer Research 57:1835-1840.

Elkind MM. 1994. Radon-induced cancer: a cell-based model of tumorgenesis due to protracted exposures. International Journal of Radiation Biology 66:649-653.

Elledge RM, Lee WH. 1995. Life and death by p53. BioEssays 17:923-930.

Ellis KJ, Cohn SH, Susskind H, Atkins HL. 1977. Kinetics of inhaled krypton in man. Health Physics 33:515-521.

Enari M, Sakahira H, Yokohama K, Iwamatsu A, Nagata S. 1998. A caspase-activated DNase that degrades DNA during apoptosis, and its inhibitor ICAD. Nature 391:43-50.

Engel E, Peskoff A, Kauffman Jr GL, Grossman MI. 1984. Analysis of hydrogen ion concentration in the gastric gel mucus layer. American Journal of Physiology 247(4):321-328.

Enoch T, Norbury C. 1995. Cellular responses to DNA damage: cell-cycle checkpoints, aptosis and the roles of p53 and ATM. Trends in Biomedical Sciences 20:426-430.

EPA. (Environmental Protection Agency). 1987a. Radiation protection guidance to federal agencies for occupational exposure. Federal Register 52(17):2822-2834.

EPA. (Environmental Protection Agency). 1987b. Technologies and Costs for Removal of Radon from Potable Water Supplies. Fourth draft. Cincinnati, OH: Environmental Protection Agency, Office of Drinking Water.

EPA. (Environmental Protection Agency). 1988a. Aeration Alternatives for Radon Reduction-Addendum to Technologies and Costs for the Removal of Radon from Potable Water Supplies. Cincinnati, OH: Environmental Protection Agency, Office of Drinking Water.

EPA. (Environmental Protection Agency). 1988b. Memorandum to Stephen Clark from Warren Peters and Chris Nelson regarding preliminary risk assessment for radon emissions from drinking water treatment facilities. Washington, DC: Environmental Protection Agency.

EPA. (Environmental Protection Agency). 1989. Memorandum to Greg Helms from Mark Parrotta Regarding Analysis of Potential Radon Emissions from Water Treatment Plants Using the MINEDOSE Code. Washington, DC: Environmental Protection Agency, Office of Drinking Water.

EPA. (Environmental Protection Agency). 1990. Suggested Guidelines for the Disposal of Drinking Water Treatment Wastes Containing Naturally Occurring Radionuclides (Draft). Washington, DC: Environmental Protection Agency, Office of Drinking Water.

EPA. (Environmental Protection Agency). 1991a. Final Draft for the Drinking Water Criteria Document. TR-1242-86. Washington, DC: US Environmental Protection Agency.

EPA. (Environmental Protection Agency). 1991b. National Primary Drinking Water Regulations: Radionuclides: Proposed Rule. 40 CFR Parts 141 and 142. Federal Register (18 July 1991), 56(138):33050-33127.

EPA. (Environmental Protection Agency). 1992a. A Citizen's Guide to Radon. 2nd Edition, EPA 402-K92-001. Washington, DC: US Environmental Protection Agency, US Department of Health and Human Services, Public Health Service.

EPA. (Environmental Protection Agency). 1992b. Groundwater disinfection rules, Part II. Federal Register 56:33960.

EPA. (Environmental Protection Agency). 1992c. National primary drinking water regulations. (Draft) Groundwater disinfection rules available for public comment. Federal Register 56:33960.

EPA. (Environmental Protection Agency). 1992d. Technical Support Document for the 1992 Citizens Guide to Radon. EPA 400-R-92-011. Washington, DC: Environmental Protection Agency, Office of Radiation and Indoor Air.

EPA. (Environmental Protection Agency). 1993a. EPA's Map of Radon Zones (State): Air and Radiation Report. 402-R-93-021 through 402-R-93-071. Washington, DC: Environmental Protection Agency.

EPA. (Environmental Protection Agency). 1994a. Model Standards and Techniques for Control of Radon in New Residential Buildings. EPA 402-R-94-009. Washington, DC: Environmental Protection Agency, Office of Air and Radiation.

EPA. (Environmental Protection Agency). 1994b. Report to the United States Congress on Radon in Drinking Water, Multimedia Risk and Cost Assessment of Radon. EPA 811-R-94-001. Washington, DC: Environmental Protection Agency, Office of Drinking Water.

EPA. (Environmental Protection Agency). 1994c. Suggested Guidelines for Disposal of Drinking Water Treatment Wastes Containing Radioactivity. Washington, DC: Environmental Protection Agency, Office of Ground Water and Drinking Water.

EPA. (Environmental Protection Agency). 1995. Uncertainty Analysis of Risk Associated with Exposure to Radon in Drinking Water. EPA 822-R-96-005. Washington, DC: Environmental Protection Agency.

EPA. (Environmental Protection Agency). 1998. Health Risks from Low-Level Environmental Exposure to Radionuclides, Federal Guidance Report No. 13, Part 1-Interim Version (Eckerman KF, Leggett RW, Nelson CB, Puskin JS, Richardson ACB). EPA 402-R-97-014. Oak Ridge National Laboratory and Environmental Protection Agency.

EPA-SAB. (Environmental Protection Agency). 1993a. An SAB Report: Multi-Media Risk Assessment for Radon. EPA-SAB-RAC-93-014. Washington, DC: Environmental Protection Agency, Science Advisory Board.

EPA-SAB. (Environmental Protection Agency). 1993b. An SAB Report: Review of Issues Related to the Cost of Mitigating Indoor Radon Resulting from Drinking Water. EPA-SAB-DWC-93-015. Washington, DC: Environmental Protection Agency, Science Advisory Board.

EPA-SAB. (Environmental Protection Agency). 1995. Recommendations for Radon Research: A Report of the Radon Science Initiative Subcommittee of the Radiation Advisory Committee. EPA-SAB-RAC-95-011. Washington, DC: Environmental Protection Agency, Science Advisory Board.

Epstein Jr EH. 1996. The genetics of human skin diseases. Current Opinion in Genetics and Development 6:295-300.

Ershow AG, Cantor FP. 1989. Total Water and Tapwater Intake in the United States: Population-Based Estimates of Quantiles and Sources. A report prepared under National Cancer Institute Order #263-MD-810264. Bethesda, MD: Federation of American Societies for Experimental Biology, Life Sciences Research Office.

Evan GI, Wyllie AH, Gilbert CS, Littlewood TD, Land H, Brookes M, Waters CM, Penn LZ, Hancock DC. 1992. Induction of apoptosis in fibroblasts by c-myc protein. Cell 69:119-128.

Fanidi A, Harrington EA, Evan GI. 1992. Cooperative interaction between c-myc and bcl-2 proto-oncogenes. Nature 359:554-556.

Fearon ER. 1997. Human cancer syndromes: clues to the origin and the nature of cancer. Nature 278:1043-1049.

Field RW, Fisher EL, Valentine RL, Kross BC. 1995. Radium-bearing pipe scale deposits: Implications for national waterborne radon sampling methods. American Journal of Public Health 85:567-570.

Fisenne IM. 1988. Radon-222 Measurements at Chester, NJ Through July 1986, 1985-1986 Biennial Report of the EML Regional Baseline Station at Chester, NJ. EML-504. New York: US Department of Energy, Enviromental Measurements Laboratory.

Fisher EL, Fuortes LJ, Field RW. 1996. Occupational exposure of water-plant operators to high concentrations of radon-222 gas. Journal of Occupational and Environmental Medicine 38:759-764.

Fisher EL, Fuortes LJ, Ledolter J, Steck DJ, Field RW. 1998. Temporal and spatial variation of waterborne point-of-use radon-222 in three water distribution systems. Health Physics 74:242-248.

Fisk WJ, Turiel I. 1983. Residential air-to-air heat exchangers: Performance, energy savings and economics. Energy and Buildings 5:197-211.

Fisk WJ, Prill RJ, Wooley J, Bonnefous YC, Gadgil AJ, Riley WJ. 1995. New methods of energy efficient radon mitigation. Health Physics 68:689-698.

Fitzgerald B, Hopke PK, Datye V, Raunemaa T, Kuuspalo K. 1997. Experimental assessment of the short and long term effects of Rn-222 from domestic shower water on the dose burden incurred in normally occupied homes. Environmental Science and Technology 31:1822-1829.

Folkman J. 1996. Tumor angiogenesis and tissue factor. Nature Medicine 2:167-168.

Fortmann RC. 1994. Measurement Methods and Instrumentation. In: Nagda, NL (ed.) Radon: Prevalence Measurements, Health Risks and Control. Philadelphia, PA: American Society for Testing and Materials. pp. 49-66.

Fowler CS, McDonough SE, Williamson AD, Sanchez DC. 1994. Passive radon control feature effectiveness in new house construction in South Central Florida. In: 1994 International Radon Symposium. Atlantic City, NJ: American Association of Radon Scientists and Technologies. pp. V-3.1 to V-3.10.

Fuscoe JC, Zimmerman LJ, Fekete A, Setzer RW, Rossiter BJ. 1992. Analysis of X-ray-induced HPRT mutations in CHO cells: insertion and deletions. Mutation Research 269:171-183.

Gadgil AJ, Bonnefous YC, Fisk WJ. 1994. Relative effectiveness of sub-slab pressurization and depressurization systems for indoor radon mitigation: Studies with an experimentally verified numerical model. Indoor Air 4:265-275.

Gadsby KJ, Harrje DT. 1991. Assessment Protocols: Durability of Performance of a Home Radon Reduction System. EPA/625/6-91/032. Research Triangle Park, NC: EPA Office of Research and Development.

George AC. 1993. Radon particle growth in the simulated environment of the respiratory tract. Aerosol Science and Technology 19:351-361.

Gesell TF, Prichard HM. 1980. The contribution of radon in tap water to indoor radon concentrations. In: Gesell TF, Lowder WM (eds.), Proceedings of the symposium on the national radiation environment III. Springfield, VA: National Technical Information Service. pp. 1347-1363.

Gesell TF. 1983. Background atmospheric ^{222}Rn concentrations outdoors and indoors: a review. Health Physics 45(2):289-302.

Getts RC, Stamato TD. 1994. Absence of Ku-like DNA end binding activity in the xrs double-strand DNA repair-deficient mutant. Journal of Biological Chemistry 269:15981-15984.

Gilroy DG, Kaschak WM. 1990. Testing of Indoor Radon Reduction Techniques in Nineteen Maryland Houses. EPA 600/8-90-056. Research Triangle Park, NC: Environmental Protection Agency, Office of Research and Development.

Godwin AR, Bollag RJ, Christie DM, Liskay RM. 1994. Spontaneous and restriction enzyme-induced chromosomal recombination in mammalian cells. Proceedings of the National Academy of Sciences 91:12554-12558.

Gogolak CV, Beck HL. 1980. Diurnal variations of radon daughter concentrations in the lower atmosphere. In: Gesell TF, Lowder WM (eds.), Natural Radiation Environment III. Report No. CONF-780422. Springfield, VA: National Technical Information Service: Department of Energy. pp. 259-280.

Graeber TG, Peterson JF, Tsai M, Monica K, Fornace Jr AJ, Giaccia AJ. 1994. Hypoxia induces accumulation of p53 protein, but activation of a GI-phase checkpoint by low-oxygen conditions is independent of p53 status. Molecular and Cellular Biology 14:6264-6277.

Graese SL, Snoeyink VL, Lee RG. 1987. Granular activated carbon filter adsorber systems. Journal of the American Water Works Association 79:64-74.

Greenblatt MS, Bennett WP, Hollstein M, Harris CC. 1994. Mutations in the p53 tumor suppressor gene: clues to cancer etiology and molecular pathogenesis. Cancer Research 54:4855-4878.

Guillouf CX, Grana M, Selvakumaran A, Deluca A, Giordana B, Hoffman B, Lieberman DA. 1995. Dissection of the genetic programs of p53-mediated growth arrest and apoptosis:blocking p53-independent apoptosis unmasks GI arrest. Blood 85:333-336.

Gundersen LCS, Schumann RR, Otton JK, Owen DE, Dubiel RF, Dickinson KA. 1992. Geology of Radon in the United States. In: Gates AE, Gundersen LCS (eds.), Geologic Controls on Radon, Special Paper No. 271. Boulder, CO: Geological Society of America. pp. 1-16.

Gundersen LCS, Schumann RR, White SJ. 1993. The USGS/EPA state geologic radon potential booklets. In: Schumann RR (eds.), Geologic Radon Potential of EPA Region [1-10]. U.S. Geological Survey Open-File Report 93-292. Reston, VA: US Geological Survey. pp. 1-35 (Chapters A-J).

Haaf T, Golub EI, Reddy G, Radding CM, Ward DC. 1995. Nuclear foci of mammalian rad51 recombination protein in somatic cells after DNA damage and its localization in synaptonemal complexes. Proceedings of the National Academy of Sciences 92:2298-2302.

Hall EJ. 1994. Radiobiology for the Radiologist. 4th edition. Philadelphia, PA: J.B. Lippincott.

Hang B, Chenna A, Fraenkel-Conrat H, Singer B. 1996. An unusual mechanism for the major human apurnic/apyrmidinic (AP) endonuclease involving 5' cleavage of DNA containing a benzene-derived exocyclic adduct in the absence of an AP site. Proceedings of the National Academy of Sciences 93:13737-13741.

Harley JH, Jetter ES, Nelson N. 1958. Elimination of Rn-222 from the Body. HASL-32. New York: U.S. Atomic Energy Commission, New York Operations Office.

Harley JH. 1990. Radon is out. In: Cross FT (eds.), Indoor radon and lung cancer, reality or myth?, 29th Hanford symposium on health and the environment. Columbus, OH: Battelle Press. pp. 741-765.

Harley JH, Jetter ES, Nelson N. 1994. Elimination of Rn-222 from the body. Environment International 20:573-584.

Harley NH, Chittaporn P, Sylvester J, Roman M. 1991. Personal and home radon and gamma-ray measured in 52 dwellings. Health Physics 61:737-744.

Harley NH, Robbins ES. 1992. Rn-222 alpha dose to organs other than the lung. Radiation Protection Dosimetry 45:619-622.

Harley NH, Robbins ES. 1994. A biokinetic model for Rn-222 gas distribution and alpha dose in humans following ingestion. Environment International 20(5):605-610.

Harley NH, Chittaporn P, Meyers OA, Robbins ES. 1996. A biological model for lung-cancer risk from ^{222}Rn exposure. Environment International 22:S977-S984.

Harley NH. 1996. Radon: over- or under-regulated? In: Mossman KL, Thiemann KB (eds.), NORM/NARM: Regulation and Risk Assessment, Proceedings of the 29th Midyear Topical Meeting of the Health Physics Society. McLean, VA: Health Physics Society. pp. 39-45.

Harley NH, Cohen BS, Robbins ES. 1996. The variability in radon decay product bronchial dose. Environment International 22:S959-S964.

Harley NH, Chittaporn P. 1997. Radon from water use in an energy efficient home.

Hastie ND, Dempster M, Dunlop MG, Thompson AM, Green DK, Allshire RC. 1990. Telomere reduction in human colorectal carcinoma and with aging. Nature 346:866-868.

Hei TK, Wu LJ, Lui SX, Vannais D, Waldren CA, Randers-Pehrson G. 1997. Mutagenic effects of a single and exact number of alpha particles in mammalian cells. Proceedings of the National Academy of Sciences 94:3765-3770.

Henschel DB. 1993. Radon Reduction Techniques for Existing Detached Houses. EPA/625/R-93-011. Research Triangle Park, NC: Environmental Protection Agency, Office of Research and Development.

Henschel DB. 1994. Analysis of radon mitigation techniques used in existing U.S. houses. Radiation Protection Dosimetry 56:21-27.

Hess CT, Weiffenbach CV, Norton SA. 1982. Variations of airborne and waterbone radon-222 in houses in Maine. Environment International 8:59-66.

Hess CT, Michel J, Horton TR, Prichard HM, Coniglio WA. 1985. The occurrence of radioactivity in public water supplies in the United States. Health Physics 48(5):553-586.

Hess CT, Vietti MA, Mage DT. 1987a. Radon from drinking water—evaluation of waterborne transfer into house air. Environmental Geochemistry and Health 9:68-73.

Hess CT, Korsah JK, Einloth CJ. 1987b. Radon in Homes Due to Radon in Potable Water. In: Hopke PK (eds.), Radon and its Decay Products: Occurrence, Properties, and Health Effects. Washington, DC: American Chemical Society. pp. 30-41.

Hess CT, Vietti MA, Lachapelle EB, Guillmette JF. 1990. Radon transferred from drinking water into house air. In: Cothern CR, Rebers PA (eds.), Radon, Radium, and Uranium in Drinking Water. Chelsea, MI: Lewis Publishers. pp. 51-67.

Hickman AW, Jaramillo RJ, Lechner JF, Johnson NF. 1994. α-particle induced p53 protein expression in a rat lung epithelial cell strain. Cancer Research 54:5797-5800.

Hintenlang DE, Ward DB, Al-Ahmady KK. 1994. Field evaluations of a radon resistant construction standard proposed for residential structures in Florida. In: 1994 International Radon Symposium. Atlantic City, NJ: American Association of Radon Scientists and Technologists. pp. II-7.1 to II-7.11.

Hollstein M, Sidransky D, Vogelstein B, Harris CC. 1991. p53 mutation in human cancers. Science 253:49-53.

Hopke PK, Ramamurthi M, Li CS. 1990. Review of the Scientific Studies of Air Cleaners as a Method of Mitigating the Health Risks from Radon and its Decay Products. Trenton, NJ: New Jersey Department of Environmental Protection, Division of Science and Research.

Hopke PK, Montassier N, Wasiolek P. 1993. Evaluation of the effectiveness of several air cleaners for reducing the hazard from indoor radon progeny. Aerosol Science and Technology 19:268-278.

Hopke PK, Jensen B, Montassier N. 1994. Evaluation of several air cleaners for reducing indoor radon progeny. Journal of Aerosol Science 25:395-405.

Hopke PK, Jensen B, Li CS, Montassier N, Wasiolek P, Cavallo AJ, Gatsby K, Socolow RH, James AC. 1995a. Assessment of the exposure to and dose from radon decay products in normally occupied homes. Environmental Science and Technology 29:1359-1364.

Hopke PK, Jensen B, Sextro R, Xu M, Nematollahi M. 1995b. Evaluation of an electrostatic air cleaner for particle and radon-decay product removal from indoor air. In: Morawska L, Bolfinger ND, Maroni M (eds.), Indoor Air—An Integrated Approach. Tarrytown, NY: Elsevier Science Ltd. pp. 351-354.

Hopke PK. 1997. Studies of the Performance of Room Air Cleaners and their Optimal Reduction of Dose from Radon Energy. Final Report, Cooperative Agreement CR-820470. Research Triangle Park, NC: Environmental Protection Agency.

Hopper RD, Levy RA, Rankin RC, Boyd MA. 1991. National ambient radon study. In: Proceedings of the 1991 EPA International Symposium on Radon and Radon Reduction Technology. EPA-600/4-9. Las Vegas, NV: Environmental Protection Agency. pp. 9-79.

Horton TR. 1983. Methods and Results of EPA's Study of Radon in Drinking Water. EPA-520/5-83-027. Washington, DC: Environmental Protection Agency.

Hursh JB, Morken DA, Davis TP, Lovaas A. 1965. The fate of radon ingested by man. Health Physics 11:465-476.

IAEA. (International Atomic Energy Agency). 1989. Evaluating the Reliability of Predictions Made Using Environmental Transport Models. Safety Series No. 100. Vienna: International Atomic Energy Agency.

ICRP. (International Commission on Radiological Protection). 1975. Report of the Task Group on Reference Man. ICRP Publication 23. Oxford: Pergamon Press.

ICRP. (International Commission on Radiological Protection). 1977. Recommendations of the International Commission on Radiological Protection. ICRP Publication 26, Annals of the ICRP 1(3). Oxford: Pergamon Press.

ICRP. (International Commission on Radiological Protection). 1981. Limits for Inhalation of Radon Daughters by Workers. ICRP Publication 32, Annals of the ICRP 6(1). Oxford: Pergamon Press.

ICRP. (International Commission on Radiological Protection). 1983. Radionuclide Transformations Energy and Intensity of Emissions. ICRP Publication 38. Oxford: Pergamon Press.

ICRP. (International Commission on Radiological Protection). 1987. Lung Cancer Risk from Indoor Exposures to Radon Daughters. ICRP Publication 50, Annals of the ICRP 17(1). Oxford: Pergamon Press.

ICRP. (International Commission on Radiological Protection). 1988. Limits for Intakes by Workers: An Addendum. ICRP Publication 30, Part 4. Oxford: Pergamon Press.

ICRP. (International Commission on Radiological Protection). 1989. Age-Dependent Doses to Members of the Public from Intake of Radionuclides, Part. 1. ICRP Publication 56, Annals of the ICRP 20(2). Oxford: Pergamon Press.

ICRP. (International Commission on Radiological Protection). 1991. 1990 Recommendations of the International Commission on Radiological Protection. ICRP Publication 60, Annals of the ICRP 21(1-3). Oxford: Pergamon Press.

ICRP. (International Commission on Radiological Protection). 1993a. Age-Dependent Doses to Members of the Public from Intake of Radionuclides: Part 2, Ingestion Dose Coefficients. ICRP Publication 67, Annals of the ICRP 23(3/4). Oxford: Pergamon Press.

ICRP. (International Commission on Radiological Protection). 1993b. Protection Against Radon-222 at Home and at Work. ICRP Publication 65, Annals of the ICRP 23(2). Oxford: Pergamon Press.

ICRP. (International Commission on Radiological Protection). 1995. Age-Dependent Doses to Members of the Public from Intake of Radionuclides: Part 4, Inhalation Dose Coefficients. ICRP Publication 71, Annals of the ICRP 25(3/4). Oxford: Pergamon Press.

ICRP. (International Commission on Radiological Protection). 1996. Age-Dependent Doses to Members of the Public from Intake of Radionuclides, Part. 5. Compilation of Ingestion and Inhalation Dose Coefficients. ICRP Publication 72, Annals of the ICRP 26(1). Oxford: Pergamon Press.

Iliakis G, Metzher L, Muschel RJ, McKenna WG. 1990. Induction and repair of DNA double strand breaks in radiation-resistant cells obtained by transformation of primary rat embryo cells with the oncogenes H-ras and V-myc: Radiation Research 50:6575-6579.

Iliakis G. 1991. The role of DNA double strand breaks in ionizing radiation-induced killing of eukaryotic cells. BioEssays 13:641-648.

Jacobi W, Eisfeld K. 1980. Dose to tissues and effective dose equivalent by inhalation of radon-222. Radon and their short lived daughters. Report S-626. Munich: Gesellschaft fur Strahlen und Umweltforschung.

Jacobson MD, Evan GI. 1994. Apoptosis. Breaking the ICE. Current Biology 4:337-340.

Jeggo PA. 1990. Studies of mammalian mutants defective in rejoining double-strand breaks in DNA. Mutation Research 239:1-16.

Jeggo PA, Taccioli GE, Jackson SP. 1995. Menage a trois: Double strand break repair, V(D)J recombination and DNA-PK. BioEssays 17:949-957.

Jen J, Powell SM, Papadopolous N, Smith KJ, Hamilton SR, Vogelstein B, Kinzler KW. 1994. Molecular determinants of dysplasia in colorectal lesions. Cancer Research 54:5523-5526.

Johnson R. 1996. Why some scientists and the public do not believe in radon risks. In: Mossman KL, Thieman KB (eds.), NORM/NARM: Regulation and Risk Assessment, Proceedings of the 29th Midyear Topical Meeting of the Health Physics Society. McLean, Virginia: Health Physics Society. pp. 65-77.

Jostes RF. 1996. Genetic, cytogenetic, and carcinogenic effects of radon: a review. Mutation Research 340:125-139.

Kadhim MA, DA MacDonald, DT Goodhead, SA Lorimore, SJ Marsden, EG Wright. 1992. Transmission of chromosomal instability after plutonium alpha-particle irradiation. Nature 355(6362):738-740.

Kadhim MA, Lorimore SA, Hepburn MD, Goodhead DT, Buckle VJ, Wright EG. 1994. α-particle-induced chromosomal instability in human bone marrow cells. Lancet 344:987-988.

Kastan MB, Canman CE, Leonard CJ. 1995. p53, cell cycle control and apoptosis: Implications for cancer. Cancer Mestat. Rev. 14:3-15.

Kennedy CJ, Probart CK, Dorman SM. 1991. The relationship between radon knowledge, concern, and behavior and health values, health locus of control, and preventive health behaviors. Health Education Quarterly 18(3):319-329.

Kennedy/Jenks Consultants. 1991a. Analysis of Costs for Radon Removal from Drinking Water Systems. Report to the American Water Works Association. San Francisco, CA: Kennedy/Jenks, Consultants.

Kennedy/Jenks Consultants. 1991b. Radon Planning Study. An Impact Analysis of the Proposed Regulation for Radon in Drinking Water on Northern Nevada Water Utilities. Prepared for Washoe County Utility Division and Westpac Utilities. San Francisco, CA: Kennedy/Jenks, Consultants.

Kerangueven F, Noguchi T, Coulier F, Allione F, Wargniez V, Simony-Lafontaine M, Longy J, Jacquemier J, Sobol H, Eisinger F, Birnbaum D. 1997. Genome-wide search for loss of heterozygosity shows extensive genetic diversity of human breast carcinomas. Cancer Research 57:5469-5474.

Kety SG. 1951. The theory and application of the exchange of inert gas at the lungs and tissue. Pharmacological Reviews 3:1-41.

Kinner NE, Lessard CE, Schell GS. 1987. Radon removal from small community water supplies using granular activated carbon and low technology/low cost techniques. In: Proceedings of the Seminar on Radionuclides in Drinking Water. Annual Conference. Denver, CO: American Water Works Association. pp. 119-140.

Kinner NE, Malley JP, Clement JA, Quern PA, Schell GS. 1989. Radon Removal Techniques for Small Community Water Supplies. EPA/600/D-89/249. Cincinnati, OH: Environmental Protection Agency, Risk Reduction Engineering Laboratory.

Kinner NE, Malley JP, Clement JA. 1990. Radon Removal Using Point-of-Entry Water Treatment Techniques. EPA/600/2-90/047. Cincinnati, OH: Environmental Protection Agency, Risk Reduction Engineering Laboratory.

Kinner NE, Malley JP, Clement JA, Fox KR. 1993. Using POE techniques to remove radon. Journal of the American Water Works Association 85(6):75-86.

Kinzler KW, Vogelstein B. 1996. Lessons from hereditary colorectal cancer. Cell 87:159-170.

Kinzler KW, Vogelstein B. 1997. Gatekeepers and caretakers. Nature 386:761-763.

Kirchgessner CU, Patil CK, Evans JW, Cuomo CA, Fried LM, Carter T, Oettinger MA, Brown JM. 1995. DNA-dependent kinase (p350) as a candidate gene for the murine SCID defect. Science 267:1178-1183.

Kladder DL, Burkhart JF, Jelinek SR. 1991. A Study of Passive Radon Reduction Techniques for Single Family Residential Dwellings. Presented at the Radon Technical Exchange, June 19-20, 1991, Golden, CO: Colorado School of Mines.

Knutson EO. 1998. Modeling indoor concentrations of radon's decay products. In: Nazaroff WW, Nero AV (eds.), Radon and Its Decay Products in Indoor Air. New York: Wiley InterScience. pp. 161-202.

Kronenberg A. 1994. Radiation-induced geometric instability. International Journal of Radiation Biology 66:603-609.

Kronenberg A, Gauny S, Criddle K, Vannais D, Ueno A, Kraemer S, Waldren CA. 1995. Heavy ion mutagenesis: linear energy transfer effects and genetic linkage. Radiation and Environmental Biophysics 34:73-78.

Kumar A, Commane M, Flickinger TW, Horvath CM, Stark GR. 1997. Defective TNF-α-induced apoptosis in STAT1-null cells due to low constitute levels of caspases. Science 278:1630-1632.

Kumar S. 1995. ICE-like proteases in apoptosis. Trends in Biochemical Sciences 20:198-202.

Lane DP. 1992. p53 guardian of the genome. Nature 358:15-16.

Lane DP. 1993. A death in the life of p53. Nature 362:786-787.

Larhed AW, Artursson P, Grasjo J, Bjork E. 1997. Diffusion of drugs in native and purified gastrointestinal mucus. Journal of Pharmaceutical Sciences 86:660-665.

Lawrence EP, Wanty RB, Nyberg P. 1992. Contribution of Rn-222 in domestic water supplies to Rn-222 in indoor air in Colorado homes. Health Physics 62(2):171-177.

Lees-Miller SP, Godbout R, Chan DW, Weinfeld M, Day RS, Barron GM, Allalunis-Turner J. 1995. Absence of p350 subunit of DNA-activated protein kinase from a radiosensitive human call line. Science 267:1183-1185.

Leggett RW, Williams LR. 1991. Suggested reference values for regional blood volume in humans. Health Physics 60:139-154.

Leggett RW, Williams LR. 1995. A proposed blood circulation model for reference man. Health Physics 69:187-201.

Leonard CJ, Canman CE, Kastan MB. 1995. The role of p53 in cell-cycle control and apoptosis: Implications for cancer. Important Advances in Oncology:33-42.

Lewis C, Hopke P, Stukel JJ. 1987. The solubility of radon in selected perflurocarbon compounds and water. Industrial Engineering and Chemical Research 26:356-359.

Li CS, Hopke PK. 1991a. Characterization of radon decay products in a domestic environment. Indoor Air 1:539-561.

Li CS, Hopke PK. 1991b. Efficacy of air cleaning systems in controlling indoor radon decay products. Health Physics 61:785-797.

Li CS, Hopke PK. 1992. Air filtration and radon decay product mitigation. Indoor Air 2:84-100.

Li P, Nijhawan D, Budihardjo I, Srinivasula SM, Ahmad M, E Alnemri S, Wang X. 1997. Cytochrome c and dATP-dependent formation of Apaf-1/capase-9 complex initiates an apoptotic protease cascade. Cell 91:479-489.

Liddament MW. 1986. Air Infiltration Calculation Techniques—An Application Guide. Report AIC-AG-1-86. Coventry, Great Britain: Air Infiltration and Ventilation Centre.

Lim DS, Hasty PA. 1996. A mutation in mouse rad51 results in early embryo lethality that is suppressed by a mutation in p53. Molecular Cell Biology 16:7133-7143.

Ling CC, Weiss H, Strauss A, Endlich B, Sheh Y, Wei JX, Orazem J. 1994. Neoplastic transformation dose response of oncogene-transfected rat embryo cells by gamma rays or 6 MeV alpha particles. Radiation Research 138:79-85.

Liu K, Hayward SB, Girman JR, Moed BA, Huang F. 1991. Annual average radon concentrations in California residences. Journal of Air and Waste Management Association 41(9):1207-1212.

Livingston EH, Engel E. 1995. Modeling of the gastric gel mucus layer: application to the measured pH gradient. Journal of Clinical Gastroenterology 21(1):S120-S124.

Loeb LA. 1991. Mutator phenotype may be required for multistage carcinogenesis. Cancer Research 51:3075-3079.

Loeb LA. 1994. Microsatellite instability: Marker of a mutator phenotype in cancer. Cancer Research 54:5059-5063.

Loevinger R, Budinger TF, Watson EE. 1988. MIRD Primer for Absorbed Dose Calculations. New York, NY: Society of Nuclear Medicine.

Longtin J. 1990. Occurrence of radionuclides in drinking water. In: Cothern RC, Rebers PA (eds.), Radon, Radium, and Uranium in Drinking Water: A national study. Chelsea, MI: Lewis Publishers. pp. 97-139.

Longtin JP. 1988. Occurrence of radon, radium, and uranium in groundwater. Journal of the American Water Works Association 80(7):84-93.

Lowe SW, Schmitt EM, Smith SW, Osborne BA, Jacks T. 1993. p53 is required for radiation-induced apoptosis in mouse thymocytes. Nature 145:163-173.

Lowry JD, Brandow JE. 1981. Removal of Radon from Groundwater Supplies Using Granular Activated Carbon or Diffused Aeration. Orono, ME: University of Maine, Department of Civil Engineering.

Lowry JD, Brandow JE. 1985. Removal of radon from water supplies. Journal of Environmental Engineering (American Society of Civil Engineers) 111:511-527.

Lowry JD, Lowry SB. 1987. Modeling point-of-entry radon removal by GAC. Journal of the American Water Works Association 79:85-88.

Lowry JD, Lowry SB, Toppan WC. 1988. New developments and considerations for radon removal from water supplies. In: Proceedings of the USEPA Symposium on Radon and Radon Reduction Technologies. Denver, CO: Environmental Protection Agency. Section VIII 2.

Lowry JD, Islam A, Paralkar B, Bezbarua B, Gould T. 1990. Adsorption, retention, and desorption of Rn and its progeny on GAC. In: Proceedings of the 1990 AWWA Annual Conference. Denver, CO: American Water Works Association. pp. 1421-1432.

Lowry JD, Lowry SB, Cline JK. 1991. Radon Removal by POE GAC Systems: Design Performance and Cost. EPA/600/S2-90/049. Cincinnati, OH: Environmental Protection Agency Risk Reduction Engineering Laboratory.

Lubin JH, Boice Jr JD, Edling C, Hornung RW, Howe G, Kunz E, Kusiak RA, Morrison HI, Radford EP, Samet JM, Tirmarche M, Woodward A, Yao SX, Pierce DA. 1994. Lung Cancer and Radon: A Joint Analysis of 11 Underground Miners Studies. Publication No. 94-3644. Bethesda, MD: National Institutes of Health.

Lubin JH, Boice Jr JD, Edling C, Hornung RW, Howe G, Kunz E, Kuziak RAMorrison HI, Radford EP, Samet JM, Tirmarche M, Woodward A, Yao SX, Pierce DA. 1995. Lung cancer in radon exposed miners and estimation of risk from indoor exposure. Journal of the National Cancer Institute 87:817-827.

Lubin JH, Boice Jr JD. 1997. Lung cancer risk from residential radon: meta-analysis of eight epidemiologic studies. Journal of the National Cancer Institute 89(1):49-57.

Lubin JH, Qiao YL, Taylor PR, Yao SX, Schatzkin A, Mao BL, Rao JY Xuan XZ, Li JY. 1990. Quantitative evaluation of the radon and lung cancer association in a case control study of Chinese tin miners. Cancer Research 50(N1):174-180.

Lucas HF. 1957. Improved low-level alpha scintillation counter for radon. Review of Scientific Instrumentation 28:680.

Ludwig D, Hilborn R, Walters C. 1993. Uncertainty, resource exploitation, and conservation: lessons learned from history. Science 260:17-36.

Luebeck EG, S Curits B, Cross FT, Moolgavkar SH. 1994. Two-stage model of radon-induced malignant tumors in rats: effects of cell killing. Radiation Research 123:127-146.

Maher EF, Rudnick SN, Moeller DW. 1987. Effective removal of airborne Rn-222 decay products inside buildings. Health Physics 53:351-356.

Mainous AG, Hagen MD. 1993. Public perceptions of radon risk. Family Practice Research Journal 13(1):63-69.

Malley Jr JP, Eliason PA, Wagler JL. 1993. Point-of-entry treatment of petroleum-contaminated water supplies. Water Environment Research 65:119-128.

Marcinowski F, White SW. 1993. EPA's map of radon zones. Health Physics 64(Suppl. 6):S47.

Marcinowski F, Lucas RM, Yeager WM. 1994. National and regional distributions of airborne radon concentrations in U.S. homes. Health Physics 66(6):699-706.

Martin SG, Miller RC, Geard CR, Hall EJ. 1995. The biological effectiveness of radon-progeny alpha particles. IV. Morphological transformation of Syrian hamster embryo cells at low doses. Radiation Research 142:70-77.

Martins KL, Meyers AG. 1993. Controlling radionuclides in water treatment plants; removal, disposal and exposure minimization. In: Proceedings of the AWWA Water Quality Technology Conference. Denver, CO: American Water Works Association. pp. 1077-1109.

McFarlane GA, Munro A. 1997. Helicobacter pylori and gastric cancer. British Journal of Cancer 84:1190-1199.

McGregor RG, Gourgon LA. 1980. Radon and radon daughters in homes utilizing deep well water supplies, Halifax County, Nova Scotia. Journal of Environmental Science and Health A15:25-35.

McTigue NE, Cornwell D. 1994. The Hazardous Potential of Activated Carbons Used in Water Treatment. Denver, CO: American Water Works Association Research Foundation.

Melek M, Gellert M, van Gent DC. 1998. Rejoining of DNA by the RAG1 and RAG2 proteins. Science 280:301-303.

Meyn MS, Strasfeld L, Allen C. 1994. Testing the role of p53 in the expression of genetic instability and apoptosis in ataxia-telangiectasia. International Journal of Radiation Biology 66(Suppl.6):S141-149.

Michel J, Jordana MJ. 1987. Nationwide distribution of Ra-228, Ra-226, Rn-222, and U in groundwater. In: Graves B (eds.), Proceedings of the National Water Well Association Conference, April 7-9,1987. pp. 227-240.

Miller RC, Marino SA, Brenner DJ, Martin SG, Richards M, Randers-Pehrson G, Hall EJ. 1995. The biological effectiveness of radon-progeny alpha particles. II. Oncogenic transformation as a function of linear energy transfer. Radiation Research 142:54-60.

Miller RC, Richards M, Brenner DJ, Hall EJ, Jostes R, Hui TE, Brooks AL. 1996. The biological effectiveness of radon-progeny alpha particles.V. Comparison of oncogenic transformation by accelerator-produced monoenergetic alpha particles and by polyenergetic alpha particles from radon progeny. Radiation Research 146:75-80.

Miller RE, Randtke SJ, Hathaway LR, Denne JE. 1990. Organic carbon and THM formation potential in Kansas groundwaters. Journal of the American Water Works Association 82(3):49-62.

Mjönes L. 1997. Changes in Radon Concentrations in Water During Passage From Waterworks to Consumer. In: Kulich J, Lundgren K, Melin J. (eds.) Translation of Statens Strålskyddinstitut Report 91-15: Förändring au Halter Radon i Onsumtiumsautten under Passage fran Vuttervert till Konsument. 91-15. Stockholm, Sweden: Statens Strälskyddinstitut.

Moolgavkar SH, Luebeck G, Krewski D, Zielinski JM. 1993. Radon, cigarette smoke, and lung cancer: A reanalysis of the Colorado Plateau miners' data. Epidemiology 4:204-217.

Morales MF, Smith RE. 1944. On the theory of blood-tissue exchanges III. Circulation and inert gas exchanges at the lung with special reference to saturation. Bulletin of Mathematical Biophysics 6:144-152.

Morgan TL, Fleck EW, Poston KA, Denovan BA, Newman CN, Rossiter BJF, Miller JH. 1990. Molecular characterization of x-ray-induced mutations at the HPRT locus in plateau-phase Chinese hamster ovary cells. Mutation Research 232:171-182.

Mose DG. 1993. Waterborne Radon Concentrations in Prince William County, Virginia. Fairfax, VA: George Mason University.

Moses H, Pearson JE. 1965. Radiological Physics Division Annual Report July 1964-June 1965. ANL-7060. Chicago, IL: Argonne National Laboratory.

Munro TR. 1970a. The relative radiosensitivity of the nucleus and cytoplasm of Chinese hamster fibroblasts. Radiation Research 42:451-470.

Munro TR. 1970b. The site of the target region for radiation-induced mitotic delay in cultured mammalian cells. Radiation Research 44:747-757.

Murane DM. 1998. New house evaluation program (NEWHEP). In: 1988 Symposium on Radon and Radon Reduction Technology. Vol. 1. Research Triangle Park, NC: Environmental Protection Agency, Office of Research and Development. pp. 9.9-9.22.

Murray DM, Burmaster DE. 1995. Residential air exchange rates in the United States: empirical and estimated parametric distributions by season and climatic region. Risk Analysis 15:459-465.

Murray DM. 1997. Residential house and zone volumes in the United States: empirical and estimated parametric distributions. Risk Analysis 17:439-446.

NAHB. (National Association of Home Builders Research Center). 1994. The New Home Evaluation Program. Draft to EPA. Upper Marlboro, MD: National Association of Home Builders Research Center.

Naismith S. 1997. Durability of radon remedial actions. Radiation Protection Dosimetry 71:215-218.

Najafi FT, Lalwani L, Li WG. 1995. Radon entry control in new house construction. Health Physics 69:67-74.

Najafi FT, Lalwani L, Peng CL, Shehata H, Shanker A, Meeske M, Roessler CE, Noble JW, Hintenlang DE. 1993. New House Evaluation of Potential Building Design and Construction for the Control of Radon in Marion and Alachua Counties, Florida. Gainesville, FL: College of Engineering, University of Florida.

National Research Council. 1977. Drinking Water and Health. Washington, DC: National Academy Press.

National Research Council. 1988. Health Risks of Radon and Other Internally Deposited Alpha-Emitters (BEIR IV). Washington, DC: National Academy Press.

National Research Council. 1989. Improving Risk Communications. Washington, DC: National Academy Press.

National Research Council. 1990a. Health Effects of Exposure to Low Levels of Ionizing Radiation, (BEIR V). Washington, DC: National Academy Press.

National Research Council. 1990b. Improving Risk Communication. Washington, DC: National Academy Press.

National Research Council. 1991a. Comparative Dosimetry of Radon in Mines and Homes. Washington, DC: National Academy Press.

National Research Council. 1991b. Frontiers in Assessing Human Exposure to Environmental Toxicants. Washington, DC: National Academy Presss.

National Research Council. 1999. Health Effects of Exposures to Radon (BEIR VI). Washington, DC: National Academy Press.

Nazaroff WW, Doyle SM, Nero AV, Sextro RG. 1987. Potable water as a source of airborne Rn-222 in U.S. dwellings: a review and assessment. Health Physics 52(3):281-295.

Nazaroff WW, Moed BA, Sextro RG. 1988. Soil as a source of indoor radon: Generation, migration and entry. In: Nazaroff WW, Nero AV (eds.), Radon and Its Decay Products in Indoor Air. New York: Wiley Interscience. pp. 57-112.

Nazaroff WW. 1992. Radon transport from soil to air. Reviews of Geophysics 30:137-160.

NCHS. (National Center for Health Statistics). 1992. Vital Statistics Mortality Data, Detail 1989. NTIS order number for datafile tape: PB92-504554. Hyattsville, MD: US Department of Health and Human Services, Public Health Service, National Center for Health Statistics.

NCHS. (National Center for Health Statistics). 1993a. Vital Statistics Mortality Data, Detail 1990. NTIS order number of datafile tape: PB93-504777. Hyattsville, MD: U.S. Department of Health and Human Services, Public Health Service, National Center for Health Statistics.

NCHS. (National Center for Health Statistics). 1993b. Vital Statistics Mortality Data, Detail 1991. NTIS order number for datafile tape: PB93-506889. Hyattsville, MD: U.S. Department of Health and Human Services, Public Health Service, National Center for Health Statistics.

NCHS. (National Center for Health Statistics). 1997. United States Life Tables. Vol.1, No. 1, DHHS, PHS-98-1150-1. Washington, DC: Public Health Service.

NCRP. (National Council on Radiation Protection and Measurements). 1975. Natural Background Radiation in the United States. NCRP Report 45. Bethesda, MD: National Council on Radiation Protection and Measurements.

NCRP. (National Council on Radiation Protection and Measurements). 1984a. Evaluation of Occupational and Environmental Exposures to Radon and Radon Daughters in the United States. NCRP Report No. 78. Bethesda, MD: National Council on Radiation Protection and Measurements.

NCRP. (National Council on Radiation Protection and Measurements). 1984b. Exposures from the Uranium Series with Emphasis on Radon and its Daughters. NCRP Report No. 77. Bethesda, MD: National Council on Radiation Protection and Measurements.

NCRP. (National Council on Radiation Protection and Measurements). 1987a. Exposure of the Population in the United States and Canada from Natural Background Radiation. NCRP Report No. 94. Bethesda, MD: National Council on Radiation Protection and Measurements.

NCRP. (National Council on Radiation Protection and Measurements). 1987b. Ionizing Radiation Exposure of the Population of the United States. NCRP Report No. 93. Bethesda, MD: National Council on Radiation Protection and Measurements.

NCRP. (National Council on Radiation Protection and Measurements). 1988. Measurement of Radon and Radon Daughters in Air. NCRP Report No. 97. Bethesda, MD: National Council on Radiation Protection and Measurements.

Nelson SL, CR Giver, AJ Grosovsky. 1994. Spectrum of X-ray-induced mutations in the human *hprt* gene. Carcinogenesis 15:495-502.

Neuberger JS. 1989. Worldwide Studies of Household Radon Exposure and Lung Cancer. Final Report to the U.S. Department of Energy, Office of Health and Environmental Research. Washington, DC: Department of Energy.

Neuberger JS, Harley NH, Kross BC. 1996. Residental radon exposure and lung cancer potential for pooled or meta analysis. Journal of Clean Technology, Environmental Toxicology and Occupational Medicine 5:207-221.

Newsham LF, Hadjistilianou T, Cavanee WK. 1998. Retinoblastoma. In: Vogelstein B, Kinzler KW (eds.), The Genetic Basis of Human Cancer. New York, NY: McGraw-Hill.

Nielson KK, Rogers VC, Holt RB. 1994. Active and passive radon control feature effectiveness estimated for the Florida residential construction standard. In: 1994 International Radon Symposium. Atlantic City, NJ: American Association of Radon Scientists and Technologists. pp. V-7.1 to V-7.9.

NIH. (National Institutes of Health). 1994. Radon and Lung Cancer Risk. A Joint Analysis of 11 Underground Miners Studies. 94-3644. Bethesda, MD: National Institutes of Health.

NIH. (National Institutes of Health). 1996. Changes in Cigarette-Related Disease Risks and Their Implication for Prevention and Control. Smoking and Tobacco Control Monograph 8. Bethesda, MD: National Institutes of Health.

Nomura A. 1996. Stomach cancer. In: Schottenfeld D, Fraumeni J. F. Jr. (eds.), Cancer Epidemiology and Prevention. 2nd Ed. New York, NY: Oxford University Press.

Nussbaum E. 1957. Radon Solubility in Body Tissues and in Fatty Acids. Report UR-503. Rochester, NY: University of Rochester.

NYDH. (New York Department of Health). 1997a. New York State Mitigation Survey. Albany, NY: New York State Department of Health.

NYDH. (New York Department of Health). 1997b. New York State Radon Awareness, Testing and Remediation Survey. Albany, NY: Department of Health.

Page S. 1993. EPA's Strategy to Reduce Risk of Radon. Journal of Environmental Health 56:27-36.

Palazzi E, Rovatti M, del Borghi M, Peloso A, Zattoni J. 1983. N-compartment mathematical model for the uptake and distribution of inhaled gases in the human body: an analytical solution. Medical and Biological Engineering and Computing 21:128-133.

Park MS, Hanks T, Jaberaboansari A, Chen DJ. 1995. Molecular analysis of gamma-ray induced mutations at the hprt locus in primary human skin fibroblasts by multiplex polymerase chain reaction. Radiation Research 141:11-18.

Pearson JE, Jones GE. 1965. Emanation of radon-222 from soils and its use as an atmospheric tracer. Journal of Geophysical Research 70:52-79.

Pearson JE, Jones GE. 1966. Soil concentrations of "emanating radon-222" and the emanation of radon-222 from soils and plants. Tellus 18:655.

Pershagen G, Ackerblom G, Axelson G, Calvensjo B, Damber L, Desai G, Enflo A, LaGarde F, Mellander H, Svatengren M, Swedjemark GA. 1994. Residential radon exposure and lung cancer in Sweden. New England Journal of Medicine 330:159-164.

Peterman BF, Perkins CJ. 1988. Dynamics of radioactive chemically inert gases in the human body. Radiation Protection Dosimetry 22:5-12.

Phillips JW, Morgan WF. 1994. Illegitimate recombination induced by DNA double-strand breaks in a mammalian chromosome. Molecular and cellular biology 14:5794-5803.

Pierce DA, Shimizu Y, Preston DL, Vaeth M, Mabuchi K. 1996. Studies of the mortality of atomic bomb survivors. Report 12, Part I. Cancer: 1950-1990. Radiation Research 146:1-27.

Pontius FW. 1998. Radon standards: problems remediating below ambient levels. In: Proceedings of the University of California at Davis Groundwater Resources Conference. September 15-17, 1997. Sacramento, CA. Davis, CA: University of California (in press).

Porstendörfer J, Röbig G, Ahmed A. 1979. Experimental determination of the attachment coefficient of atoms and ions on monodisperse aerosols. Journal of Aerosol Science 10:21-28.

Pourzand C, Rossier G, Reelfs O, Borner C, Tyrell RM. 1997. The overexpression of Bcl-2 inhibits UVA-mediated immediate apoptosis in rat 6 fibroblasts: evidence for the involvement of Bcl-2 as an antioxidant. Cancer Research 57:1405-1411.

Price JG, Rigby JG, Christensen L, Hess R, LaPointe DD, Ramelli AR, Desilets M, Hopper RD, Kluesner T, Marshall S. 1994. Radon in outdoor air in Nevada. Health Physics 66(4):433-438.

Price PN. 1997. Prediction and maps of county mean indoor radon concentrations in the mid-atlantic states. Health Physics 72(6):893-906.

Prill RJ, Fisk WJ, Turk BH. 1990. Evaluation of radon mitigation systems in 14 houses over a two-year period. Journal of the Air and Waste Management Association 40:740-746.

Raabe OG. 1987. Three-dimensional dose response models of competing risks and natural lifespan. Fundamental Applied Toxicology 8:463-473.

Rabbitts TH. 1994. Chromosomal translocations in human cancer. Nature 372:143-149.

Rai SN, Krewski D. 1998. Uncertainty and variability analysis in multiplicative risk models. Risk Analysis 18:37-45.

Ramamurthi M, Hopke PK. 1989. Improving the validity of wire screen unattached fraction Rn daughter measurements. Health Physics 56:189-194.

Ramamurthi M, Hopke PK. 1991. An automated, semi-continuous system for measuring indoor radon progeny activity-weighted size distributions, dp: 0.5-500nm. Aerosol Science and Technology 14:82-92.

Rand PW, Lacombe EH, Perkins WD Jr. 1991. Radon in homes following reduction in a community water supply. Journal of the American Water Works Association 83:154-158.

Rathmell WK, Chu G. 1994. Involvement of the Ku autoantigen in the cellular response to DNA double-strand breaks. Proceedings of the National Academy of Sciences 91:7263-7627.

Reed BE, Arunachalam S. 1994. Use of granular activated carbon columns for lead removal. Journal of Environmental Engineering (American Society of Civil Engineers) 120:416-436.

Reed JC. 1997. Cytochrome C: can't live with it, can't live without it. Cell 91:559-562.

Revzan KL, Fisk WJ. 1992. Modeling radon entry into houses with basements: the influence of structural factors. Indoor Air 2:40-48.

Riley WJ, Gadgil AJ, Nazaroff WW. 1996. Wind-induced ground-surface pressures around a single-family house. Journal of Wind Engineering and Industrial Aerodynamics 61:153-167.

Ritter MA, Cleaver JE, Tobias CA. 1977. High LET radiations induce a large proportion of non-rejoining DNA breaks. Nature 266:653-655.

Robinson AL, Sextro RG. 1995. The influence of a subslab gravel layer and open area on soil-gas and radon entry into two experimental basements. Health Physics 69:367-377.

Robinson AL, Sextro RG. 1997. Radon entry into buildings driven by atmospheric pressure fluctuations. Environmental Science and Technology 31:1742-1748.

Rogers VC, Nielson KK. 1994. Incorporating a radon potential map into the Florida residential construction standard. In: 1994 International Radon Symposium. Atlantic City, NJ: American Association of Radon Scientists and Technologists. pp. VP-3.1 to VP-3.8.

Rubin AJ, Mercer DL. 1981. Adsorption of Free and Complexed Metals from Solution by Activated Carbon; Adsorption of Inorganics at Solid Liquid Interfaces. Ann Arbor, MI: Ann Arbor Science.

Rudnick SN, Hinds WC, Maher EF, First MW. 1983. Effect of plateout, air motion and dust removal on radon decay product concentration in a simulated residence. Health Physics 45:463-470.

Rydell S, Keene B, Lowry J. 1989. Granular activated carbon water treatment and potential radiation hazards. Journal of the New England Water Works Association 103:234-248.

Rydell S, Keene B. 1993. CARBDOSE Version 3.0. Boston, MA: US Environmental Protection Agency, Region I.

Saccomanno G, Auerbach O, Kuschner M, Harley NH, Michels RV, Anderson MW, Betchel JJ. 1996. A comparison between the localization of lung tumors in uranium miners and in nonminers from 1947-1991. Cancer 77(7):1278-1283.

Sachs HM, Hernandez TL, Ring JW. 1982. Regional geology and radon variability in buildings. Environment International 8:97-103.

Sancar A, Sancar GB. 1988. DNA repair enzymes. Annual Review of Biochemistry 57:29-67.

Sanchez DC, Dixon R, Williamson AD. 1990. The Florida radon research program: Systematic development of a basis for statewide standards. In: The 1990 International Symposium on Radon and Radon Reduction Technology. Vol. 1 (Preprints). Research Triangle Park, NC: Environmental Protection Agency, Office of Research and Development. Paper A-I-3.

Sandman PM. 1993. Predictors of home radon testing and implications for testing promotion programs. Health Education Quarterly 20(4):471-487.

Sankaranarayanan K. 1991. Ionizing radiation and genetic risks: III Nature of spontaneous and radiation-induced mutations in mammalian in vitro system and mechanism of induction of mutations by radiation. Mutation Research 258:75-97.

Sargent RG, Brenneman MA, Wilson JH. 1997. Repair of site-specific double-breaks in a mammalian chromosome by homologous and illegitimate recombination. Molecular and Cellular Biology 17:267-277.

Saum DW, Osborne MC. 1990. Radon mitigation performance of passive stacks in residential new construction. In: 1990 International Symposium on Radon and Radon Reduction Technology. Vol. 5. Research Triangle Park, NC: U.S. Environmental Protection Agency, Office of Research and Development. Paper VIII-2.

Saum DW. 1991. Mini fan for SSD radon mitigation in new construction. In: 1991 International Symposium on Radon and Radon Reduction Technology. Vol. 4. Research Triangle Park, NC: Environmental Protection Agency, Office of Research and Development. Paper VIII 5.

Schumann RR, Owen EO, Asher-Bolinder SA. 1992. Effects of weather and soil characteristics on the temporal variations in soil-gas radon concentrations. In: Gates AE, Gunderson LCS (eds.), Geologic Controls on Radon. Geological Society of America Special Paper 271. Boulder, CO: Geological Society of America. pp. 65-72.

Schumann RR, Gundersen LCS, Tanner AB. 1994. Geology and occurrence of radon. In: Nagda NL (eds.), Radon: Prevalence, Measurement, Health Risks and Control. Philadelphia, PA: American Society for Testing and Materials. pp. 83-96.

Scully R, Chen J, Plug A, Xiao Y, Weaver D, Feunteun J, Ashley T, Livingston DM. 1997a. Association of BRCA1 with rad51 in mitotic and meiotic cells. Cell 88:265-275.

Scully R, Chen J, Ochs RL, Keegan K, Hoekstra M, Fuenten J, Livingston DM. 1997b. Dynamic changes of BRCA1 subnuclear location and phosphorylation state are initiated by DNA damage. Cell 90:425-435.

Servant J. 1966. Temporal and spatial variations of the concentration of the short-lived decay products of radon in the lower atmosphere. Tellus 18:663-671.

Sextro RG, Offermann FJ, Nazaroff WW, Nero AV, Revzan KL, Yater J. 1986. Evaluation of indoor aerosol control devices and their effects on radon progeny concentrations. Environment International 12:429-438.

Sextro RG. 1994. Radon and the natural environment. In: Nagda NL (eds.). Radon: Prevalence, Measurements, Health Risks and Control. Philadelphia, PA: American Society for Testing and Materials. pp. 9-32.

Shah GM, Shah RG, Poirier GG. 1996. Different cleavage pattern for poly(ADP-ribose) polymerase during necrosis and apoptosis in HL-60 cells. Biochemical Biophysical Research Communications 229:838-844.

Sharma N, Hess CT, Thrall KD. 1996. A compartmental model of water radon contamination in the human body. Health Physics 72(2):261-268.

Sherman M, Matson N. 1997. Residential ventilation and energy characteristics. ASHRAE Transactions 103(1):717-730.

Sherman MH, Dickerhoff DJ. 1994. Air Tightness of U.S. Dwellings, Proceedings of the 15th AIVC Conference. LBL-35700. Berkeley, CA: Lawrence Berkeley Laboratory.

Sikov MR, Cross FT, Mast TJ, Palmer ME, James AC, Thrall KD. 1992. Developmental toxicology of radon exposures. In: Cross FT (eds.), Indoor Radon and Lung Cancer: Reality or Myth? 29th Hanford Symposium on Health and the Environment. Columbus, OH: Battelle Press. pp. 677-691.

Singer B. 1996. DNA: damage chemistry, repair, and mutagenic potential. Regulatory Toxicology and Pharmacology 1(Pt. 1):2-13.

Singer B, Hang B. 1997. What structural features determine repair enzyme specifity and mechanism in chemically modified DNA? Chemical Research in Toxicity 10:713-732.

Smider V, Rathmell WK, Lieber MR, Chu G. 1994. Restoration of X-ray resistance in V(D)J recombination and mutant cells by Ku cDNA. Science 266:288-291.

Smith CL. 1964. Microbeam and partial cell irradiation. International Review of Cytology 16:133-153.

Smith RE, Morales MF. 1944. On the theory of blood-tissue exchanges: I. Fundamental equations. Bulletin of Mathematical Biophysics 6:125-131.

Sonoda E, Sasaki MS, Buerstedde J-M, Bezzubova O, Shinohara A, Ogawa H, Takata Y, Yamaguchi-Iwai Y, Takeda S. 1998. Rad51-deficient vertebrate cells accumulate chromosomal breaks prior to cell death. The EMBO Journal 17(2):598-608.

Sorg TJ, Logsdon GS. 1978. Treatment technology to meet interim primary drinking water regulations for inorganics: Part 2. Journal of the American Water Works Association 70:379-392.

Sorg TJ. 1988. Methods for removing uranium from drinking water. Journal of the American Water Works Association 80:105-111.

Spears JW, Nowak MS. 1988. Radon mitigation in new construction: Four case studies. In: 1988 Symposium on Radon and Radon Reduction Technology. Vol. 1. Research Triangle Park, NC: Environmental Protection Agency, Office of Research and Development. pp. 9-23 to 9-36.

Steck DJ. 1992. Spatial and Temporal Indoor Radon Variations. Health Physics 62:351-355.

Steck DJ, Field RW, Lynch CF. 1999. Exposure to atmospheric radon. Environmental Health Perspectives 107:3-7.

Steward A, Allott PR, Cowles AL, Mapleson WW. 1973. Solubility coefficients for inhaled anaesthetics for water, oil, and biological media. British Journal of Anaesthesia 45:282-292.

Stidley CA, Samet JM. 1994. Assessment of ecologic regression in the study of lung cancer and indoor radon. American Journal of Epidemiology 139:312-322.

Stoler DL, Anderson GR, Russo CA, Spina AM, Beerman TA. 1992. Anoxia-inducible endonuclease activity as a potential basis of the genomic instability of cancer cells. Cancer Research 52:4372-4378.

Strong KP, Levins DM. 1978. Dynamic adsorption of radon on activated carbon. In: 15th DOE Nuclear Air Cleaning Conference. Las Vegas, NV: US Department of Energy. pp. 627-639.

Suomela M, Kahlos M. 1972. Studies on the elimination rate and the radiation exposure following ingestion of Rn-222 rich water. Health Physics 23:641-652.

Susskind H, Atkins HL, Cohn SH, Ellis KJ, Richards P. 1976. The kinetics of total body retention and clearance of xenon and krypton after inhalation. In: Proc. 29th Annual Conference on Engineering Medicine and Biology. pp. 225.

Susskind H, Atkins HL, Cohn SH, Ellis KJ, Richards P. 1977. Whole body retention of radioxenon. Journal of Nuclear Medicine 18:462-471.

Taccioli GE, Gottlieb TM, Blunt T, Priestley A, Demengeot J, Mizuta R, Lehmann AR, Alt FW, Jackson SP, Jeggo PA. 1994. Ku80: Product of the XRCC5 gene and its role in DNA repair and V(D)J recombination. Science 278:128-130.

Taddei F, Hayawaka H, Bouton M, Cirenesi A, Matic I, Sekiguchi M, Radman M. 1997. Counteraction by MuT protein of transcriptional errors caused by oxidative damage. Science 278:128-130.

Takata M, Sasaki MS, Sonoda E, Morrison C, Yamaguchi-Iwai Y, Takeda S. in press. Homologous recombination and nonhomologous endjoining pathways of vertebrate cells have overlapping roles in DNA double-strand breaks repair and maintenance of chromosomal integrity. The EMBO Journal.

Tamburini JU, Habenicht WL. 1992. Volunteers integral to small system's success. Journal of the American Water Works Association 84:56-61.

Tanner AB. 1964. Radon migration in the ground: a review. In: Adams JAS, Lowder WM (eds.), The Natural Radiation Environment. Chicago, IL: University of Chicago Press. pp. 161-190.

Tanner AB. 1980. Radon migration in the ground: a supplementary review. In: Gesell TF, Lowder WM (eds.), Natural Radiation Environment III. Report No. CONF-780422. Springfield, VA: National Technical Information Service. pp. 5-56.

Thaler I, Manor D, Itskovitz J, Rotten S, Levit N, Timor-Tritsch I, Brandes JM. 1990. Changes in uterine blood flow during human pregnancy. American Journal of Obstetrics and Gynecology 162:121-125.

Thomas JW. 1973. Radon adsorption by activated carbon in uranium mines. In: Proceedings of the Noble Gas Symposium. Las Vegas, NV: Environmental Protection Agency, National Environmental Research Center and University of Nevada. pp. 637-646.

Thompson LH, Fong S, Brookman K. 1980. Validation of conditions for efficient detection of HPRT and APRT mutations in suspension-cultured Chinese hamster cells. Mutation Research 74:21-36.

Thompson LH. 1996. Evidence that mammalian cells possess homologous recombinational repair pathways. Mutation Research 363:77-88.

Tlsty T. 1996. Regulation of genomic instability in preneoplastic cells. Cancer Surveys 28:217-224.

Tlsty TD, Briot A, Gualberto A, Hall I, Hess S, Hixon M, Kuppuswamy D, Romanov S, Sage M, White MA. 1995. Genomic instability and cancer. Mutation Research 337:1-7.

Tobias CA, Jones HB, Lawrence JH, Hamilton JG. 1949. The uptake and elimination of krypton and other inert gases by the human body. Journal of Clinical Investigation 28:1375-1385.

Tomasek L, Darby SC, Swerdlow AJ, Placek V, Kunz E. 1993. Radon exposure and cancers other than lung cancer among miners in west Bohemia. Lancet 341:919-923.

Tsuzuki T, Fujii Y, Sakumi K, Tominga Y, Nakao K, Sekiguchi M, Matsushiro A, Yoshimura Y, Morita T. 1996. Targeted disruption of the Rad51 gene leads to lethality in embryonic mice. Proceedings of National Academy of Sciences 93:6236-6240.

Tu KW, Knutson EO, George AC. 1991. Indoor radon progreny aerosol size measurements in urban, suburban and rural regions. Aerosol Science and Technology 15:170-178.

Tucker JD, Breneman JW, Briner JF, Evelah GG, Langlais RG, Moore DH. 1997. Persistence of radiation-induced translocations in rat peripheral blood determined by chromosome painting. Environmental and Molecular Mutagenesis 30:264-272.

Turk BH, Harrison J, Prill RJ, Sextro RG. 1990. Developing soil gas and Rn-222 entry potentials for substructure surfaces and assessing Rn-222 control diagnostic techniques. Health Physics 59:405-419.

Turk BH, Prill RJ, Fisk WJ, Grimsrud DT, Sextro RG. 1991a. Effectiveness of radon control techniques in fifteen homes. Journal of the Air and Waste Management Association 41:723-734.

Turk BH, Harrison J, Sextro RG. 1991b. Performance of radon control systems. Energy and Buildings 17:157-175.

Uckun FM, Tuel-Ahlgren L, Song CW, Waddick K, Myers DE, Kirihara J, Ledbetter JA, Schieven GL. 1992. Ionizing radiation stimulates unidentified tyrosine-specific protein kinases in human B-lymphoctye precursors, triggering apoptosis and clonogenic cell death. Proceedings of the National Academy of Sciences 89:9005-9009.

Ueno AM, Vannais DB, Gustafson DL, Wong JC, Waldren CA. 1996. A low adaptive dose of gamma rays reduced the number and altered the spectrum of S1-mutants in human-hamster hybrid A_L cells. Mutation Research 358:161-169.

Unger C, Kress C, Buchmann A, Schwarz M. 1994. Gamma-irradiation-induced micronuclei from mouse hepatoma cells accumulate high levels of p53. Cancer Research 54:3651-3655.

UNSCEAR. (United Nations Scientific Committee on the Effects of Atomic Radiation). 1988. Sources, Effects and Risks of Ionizing Radiation. New York: United Nations.

UNSCEAR. (United Nations Scientific Committee on the Effects of Atomic Radiation). 1993. Sources and Effects of Ionizing Radiation. New York: United Nations.

US Congress. 1986. Safe Drinking Water Act Amendments (SDWA). Public Law 99-939, 100 STAT 642.

US Congress. 1996. Safe Drinking Water Act Amendments (SDWA). Public Law 104-182.

Vahakangas KH, Samet JM, Metcalf RA, Welsh JA, Bennett WP, Lane DP, Harris CC. 1992. Mutations of p53 and ras genes in radon-associated lung cancer from uranium miners. Lancet 339:576-580.

Von Doebeln W, Lindell B. 1964. Some aspects of radon contamination following ingestion. Arkiv fur Fysik 27:531-572.

Waddick KG, Chae HP, Tuel-Ahlgren L, Jarvis LJ, Dibridik I, Myers DE, Uckun FM. 1993. Engagement of the CD19 receptor on human B-lineage leukemia cells activates LCK tyrosine kinase and facilitates radiation-induced apoptosis. Radiation Research 136:313-319.

Wallace LA. 1997. Human exposure and body burden for chloroform and other trihalomethanes. Critical Reviews in Environmental Science and Technology 27(2):113-194.

Ward JF. 1988. DNA damage produced by ionizing radiation in mammalian cells: Identities, mechanism of formation, and repairability. Progress in Nucleic Acid Research and Molecular Biology 35:95-125.

Ward JF, Webb CF, Limoli CL, Milligan JR. 1990. DNA lesions produced by ionizing radiation: Locally multiply damages sites. In: Wallace SS, Painter RB (eds.), Ionizing Radiation Damage to DNA: Molecular Aspects. New York: Wiley-Liss.

Ward JF. 1990. The yield of DNA double-strand breaks produced intracellularly by ionizing radiation: a review. International Journal of Radiation Biology 57:1141-1150.

Wasiolek P, Montassier N, Hopke PK, Abrams R. 1993. Analysis of the performance of a radon mitigation system based on charcoal beds. Environmental Technology 14:401-412.

Wasiolek PT, Schery SD. 1993. Outdoor radon exposure and doses in Socorro, New Mexico. Radiation Protection Dosimetry 46(1):49-54.

Wasiolek PT, Schery SD, Broestl JE. 1996. Experimental and modeling studies of Rn-220 decay products in outdoor air near the ground surface. Environment International 22(Suppl. 1):S193-S204.

Wilcox DP, Chang E, Dickson KL, Johansson KR. 1983. Microbial growth associated with granular activated carbon in a pilot water treatment facility. Applied and Environmental Microbiology 46:406-420.

Wilkening MH, Clements WE. 1975. Radon-222 from the ocean surface. Journal of Geophysical Research 80:3828.

Williams LR. 1993. Reference values for total blood volume and cardiac output in humans. ORNL/TM-12814. Oak Ridge,TN: Oak Ridge National Laboratory.

Williams LR, Leggett RW. 1989. Reference values for resting blood flow to organs of man. Clinical Physics and Physiological Measurement 10:187-217.

WSBCC. (Washington State Building Code Council). 1991. Washington State Ventilation and Indoor Air Quality Code. 2nd Edition. Chapters 51-13 WAC.

Xanthoudakis S, Smeyne RJ, Wallace JD, Curran T. 1996. The redox/DNA repair protein, Ref-1, is essential for early embryonic development in mice. Proceedings of the National Academy of Sciences 93:8919-8923.

Xie J, Murone M, Luoh SM, Ryan A, Gu Q, Zhang C, Bonifas JM, Lam CW, Hynes M, Goddard A, Rosenthal A, Epstein Jr EH, de Sauvage FJ. 1998. Activating smoothened mutations in sporadic basal-cell carcinoma. Nature 391:90-92.

Xuan, X-Z, Lubin JH, Jun-Yao L, Li-Fen Y, Sheng LQ, Lan Y, Jian-Zhang W, Blot WJ. 1993. A cohort study in southern China of tin miners exposed to radon and radon decay products. Health Physics 64(2):120-131.

Zhang Y, Woloschak GE. 1997. Rb and p53 gene deletions in lung adenocarcinomas from irradiated and control mice. Radiation Research 148:81-89.

Zhu LX, Waldren CA, Vannias D, Hei TK. 1996. Cellular and molecular analysis of mutagenesis by charged particles of defined linear energy transfer. Radiation Research 145:251-259.

Glossary

Absolute risk. An expression of risk based on the assumption that the excess risk from exposure to ionizing radiation is *added* to the underlying (baseline) risk by an increment that is dependent on dose but independent of the underlying baseline risk.

Absorbed dose (D). The mean energy absorbed in material from any type of ionizing radiation divided by the mass of the material. Absorbed dose, D, has the dimensions of energy divided by mass and is expressed in gray (Gy) or rad.

Accuracy. The tendency of measurements (estimates) of a quantity to yield, on average, the true value of the quantity. Accuracy is the complement of bias.

Active subslab depressurization. Mechanically assisted method, e.g., fan or suction device, to remove soil gas from beneath the foundation of a building.

Advection. Bulk flow of gas due to temperature or pressure differences.

Aerodynamic diameter. The descriptive size of any type of aerosol based on the diameter of a sphere of water that has the same settling velocity as the aerosol of interest.

Alpha particle (α). A particle emitted during the decay of certain radioactive elements. It is identical to the nucleus of helium containing two protons and two neutrons.

Alternative maximum contaminant level (AMCL). For the case of radon in water, the AMCL is the concentration in water that will contribute an incremental increase in indoor air concentration equal to the national average ambient-air concentration.

Angiogenesis. The process of developing new blood vessels, especially for growing tissues or tumors.

Apoptosis. A process of degradation by proteases (caspases) and nucleases which results in a noninflammatory mechanism of cell destruction and resorption.

Apurinic site. A site in DNA that remains after a damaged base has been removed by an enzyme that cleaves the deoxyribose-base bond.

Attributable risk (AR). The proportion of excess cancer deaths in a defined population that could, in theory, be prevented if all exposures to radon were reduced to background concentrations.

Backwashing. The process used to clean a filter bed after the flow of water being treated is reduced due to clogging.

Bayesian. A statistical methodology that allows for the incorporation of prior information and the use of subjective probability.

223

Becquerel (Bq). A quantity of radioactivity equivalent to 1 decay per second. The Becquerel is a SI unit named on behalf of the French scientist Henri Becquerel.

BEIR IV. The fourth in a series of National Research Council reports called the Biological Effects of Ionizing Radiation. The 1988 report is titled: *Health Risks of Radon and Other Internally Deposited Alpha-Emitters.*

BEIR VI. The sixth in a series of National Research Council reports called the Biological Effects of Ionizing Radiation. The 1999 report is titled: *Health Risks of Exposure to Radon.*

Best available technology (BAT). The most efficient treatment method for removal of a given contaminant.

Beta particle (β). A particle emitted during the decay of certain radioactive elements. It is identical to an electron.

Bias. The difference between the average value obtained from a measurement (estimate) of a specific quantity and the true value of that quantity. Bias is the complement of accuracy.

Cancer. A malignant tumor of potentially unlimited growth, capable of invading surrounding tissue or spreading to other parts of the body by metastasis.

Carcinogen. An agent that can cause cancer.

Caspase. A family of proteolytic enzymes that are normally found complexed with peptide inhibitors which are released by caspase activity itself. This results in an autocatalytic increase in protease activity (cascade) that degrades specific protein substrates especially those involved in structural components of the cell.

Chromatin. The combination of DNA and proteins which together make up the main structural units of the nucleus and the chromosomes. The first order of structure consists of DNA wrapped around specific histone proteins packed together to form nucleosomes.

Countercurrent flow. The hydraulic regime in a treatment unit where the flow of the fluid being cleaned is in the opposite direction to that of the fluid to which the contaminant is transferred.

Crypts. Pits formed by depressions in the surface of the stomach or intestinal lining cells within the crypts divide and secrete digestive enzymes.

Curie (Ci). A quantity of radioactivity equivalent to 3.7×10^{10} (37 billion) decays per second. The curie is a traditional unit named on behalf of the French scientist Marie Curie.

Cytogenetics. The branch of genetics devoted to the study of chromosomes.

Cytokines. Extracellular molecules that transmit signals to control gene expression in target cells, often through interaction with specific receptors on the cell surface.

Diploid. The normal state of the genome of cells in most body tissues in which each cell has two copies of each chromosome, one copy originating from each parent.

Disinfection by-products (DBPs). Compounds formed when organic matter in raw water is oxidized by disinfectants. For example, chloroform is formed when natural organic matter is oxidized by chlorine.

DNA homologs. Two DNA sequences or chromosomal regions which have sufficiently similar nucleotide sequence to represent the same genes or intervening sequences.

Effective dose (H_E). The product of the equivalent dose, H_T, in a tissue, T, multiplied by the tissue weighting factor w_T. The purpose of this is to take into consideration the difference in sensitivity of various tissues or organs to radiation induced cancer.

Electron volt (eV). A unit of energy equal to the kinetic energy gained by a particle having one electronic charge when it passes in a vacuum through a potential difference of 1 volt. $1 \text{ eV} = 1.602 \times 10^{19}$ joules. 1 MeV = 1,000,000 eV. 1 keV = 1,000 eV.

Empty bed contact time (EBCT). The average interval of time that a fluid containing a contaminant remains in contact with a bed of granular activated carbon. EBCT is the volume of the reactor (not containing GAC) divided by the flow rate of water being treated.

Endogenous. A term describing sources of reactive molecules that originate from within the cell.

Episome. A small circular DNA molecule that can be maintained for varying periods of time within the nucleus but is not a functional part of the DNA for the host cell.

Epithelial tissue. The cell types on the outer surface of mammalian tissues. Endothelia cell layers are on the inner surfaces of tissues. Cells in-between are mesothelial.

Equilibrium ratio (F). The ratio of the potential alpha energy concentration of radon decay products to the concentration of radon. Under ideal conditions, when all of the radon decay products remain suspended in air, they reach equilibrium with radon and F approaches 1.0. In family dwellings, F is typically near 0.4.

Equilibrium-equivalent radon concentration (EEC). The concentration of radon in equilibrium with the short-lived decay products that has the same potential alpha concentration as a given mixture of decay products.

Equivalent dose (H). The product of absorbed dose, D, and a radiation weighting factor, w_R, that depends on the type of radiation responsible for the dose. $H = w_R \cdot D$. The purpose is to account for the differences in biological response for different types of radiation. If dose is measured in Gy, H has units of Sievert (Sv). If D is measured in rad, H has units of rem.

Excess relative risk (ERR). A model, which assumes that health effects from ionizing radiation are based on a relative risk factor, RR, that *multiplies* the baseline risk. Excess relative risk, ERR, is then defined formally as: RR-1.

Exogenous. A source of exposure that is outside the body.

Fluvial. Pertaining to, produced by, or formed in a stream or river.

Fos/jun. Two oncogenes that interact as a dimer which binds to specific DNA control sequences regulating the transcription of genes. Fos/jun is particularly responsive to DNA damage and induces transcription of damage-responsive genes.

Fusion gene. A gene produced by breakage and rejoining of DNA within gene sequences to produce a new gene with altered function. Bcl2 is involved in a well-known fusion gene produced during some classes of hematopoetic cancer.

G_1 and G_2. The periods in the cell cycle that are respectively before and after, the period of semiconservative DNA synthesis called the S phase. In the process of cell division, mitosis, follows the completion of G_2. Most cells in somatic tissues are either in the G_1 phase of the mitotic cycle, or are not in a mitotic cycle at all and are then classified as being in G_0.

Gamma ray (γ). A particle emitted during the decay of certain radioactive elements. It is a form of electromagnetic radiation also referred to as a photon. Gamma rays have energies usually between 10 keV and 10 MeV.

Genomic stability. The concept where the genome of a normal diploid cell maintains a complement of maternal and paternal genes by specific mechanisms. These become deranged in malignant cells and can cause changes in gene copy number, chromosome numbers, heterozygosity, total DNA content, etc.

Geometric mean (GM). The nth root of the product of n observations.

Geometric Standard Deviation (GSD). The exponential of the standard deviation of the natural logarithms of a set of observations.

Granular activated carbon (GAC). Organic matter such as wood, bone or coconut shells, that is exposed to high heat and pressure. This increases the surface area and improves the capability to adsorb contaminants.

Gray (gy). A quantity of absorbed dose equal to 1 Joule kg^{-1}. One Gy = 100 rad. The gray is a SI unit named on behalf of the British scientist L.H. Gray.

Groundwater disinfection rule (GWDR). Pending rule proposed by USEPA that will require public drinking water supplies that use groundwater to disinfectant the drinking water prior to distribution.

Half-life ($t_{1/2}$). The amount of time required for a given quantity of radioactivity to be reduced by one half. This only includes the radioactive decay process and does not include removal radioactivity by other methods such as biological elimination or migration.

Hematopoeitic. The lineage of cells in the bone marrow, spleen and thymus that produces the erythrocytes, lymphocytes, platelets and other cells of the peripheral blood.

Heterotrophic plate counts (HPC). A plating method for enumerating the number of viable organic carbon-using (heterotrophic) bacteria in a water sample.

Heterozygosity. The genetic state in which the two genes specifying a particular enzyme or protein, from both parents contain DNA sequence inferences.

Homeostasis. The stable expression of total cellular metabolism.

Homologous recombination. A mechanism of DNA repair and genetic exchange between two DNA homologs.

Hormesis. The concept that very low doses of ionizing radiation may be beneficial to the irradiated cells or organisms.

ICRP. International Commission on Radiological Protection and Measurements, founded in 1928 and since 1950 has been providing general guidance on radiation protection.

Immunoglobulin. The protein consisting of pairs of heavy and light chains that make the circulating antibodies which exhibit great diversity in recognition and binding target molecules known as antigens.

Interphase. The period of the cell cycle during which chromosomes are not visible as discrete structures. At this time the nucleus of the cell is somewhat like a spherical zone surrounded by a nuclear membrane that contains the encapsulated DNA and also regulates the traffic of molecules between the inner volume of the nucleus and the outer volume of the cytoplasm. Interphase can be subdivided into G1, S, and G2 phases according to the state of duplication of the DNA.

Ion exchange treatment. A method used to remove anionic or cationic contamination from water. The contaminants adhere to locations containing anions or cations normally associated with the resin. When the resin becomes saturated with the contaminant, it is regenerated with a brine containing a high concentration of the originally-sorbed anion or cation.

Ionization. A process by which a neutral atom or molecule loses or gains electrons, thereby acquiring a net electric charge. Ionization can be produced by the interaction of ionizing radiation with matter.

Karstic. A type of topography characterized by sink-holes, caves, and underground drainage, usually formed in limestone or salt deposits.

Karyotype. The full set of chromosomes in the nucleus of a cell that is characteristic of a particular individual or species.

Lacustrine. Pertaining to, produced by, or formed in a lake.

Linear no-threshold (LNT) model. A risk-projection model that expresses the effect (e.g., mutation or cancer) as a proportional (linear) function of the dose and assumes that no minimum (threshold) dose exists below which radiation injury does not occur.

Loss of heterozygosity (LOH). A process occurring during tumor progression by which one chromosome of a pair is lost and the partner is duplicated.

Lumen. The interior open space of an organ such as a blood vessel or intestine.

Maximum contaminant level (MCL). The highest concentration of a contaminant permitted by regulations in public drinking water supplies.

Meiotic. The process by which a germ cell in the testis or ovary divides into two cells with a reduction in chromosome number such that each cell has one copy of each chromosome, and ultimately develops into either sperm or egg cells.

Meta-analysis. An analysis of epidemiological data based on grouping or pooling information obtained from several studies.

Microsatellite repeats. Regions of DNA in which the same short sequence is repeated numerous times to create characteristic sequence motifs useful for individual identification and which slow expansion and reduction in size in some tumors.

Mitochondria. Organelles in the cytoplasm of cells that contain a small circular DNA molecule which encodes many of the genes required for oxidative phosphorylation and provides most of the energy for the cell through ATP production.

Mitotic. The process by which a cell divides into two identical daughter cells with no change in chromosome number; also used to refer to the whole cell cycle during which this cell division occurs.

Mitotic cell death. The result of cells attempting to go through mitotic cell division with broken or fused chromosomes such that the daughter cells do not receive the full complement of DNA necessary for survival.

Monte Carlo (analysis, methods, simulation). A numerical technique that samples values at random from specified probability distributions.

MutT. An enzyme which hydrolyzes 8-ozyguanine triphosphates to monophosphates and eliminates them as precursors for DNA and RNA synthesis.

Necrosis. An ill-defined form of cell death that may be different from apoptosis, and is often characterized by sudden collapse of nuclear and cytoplasmic structures and loss of membrane integrity.

Neoplastic. An alternative term for malignant or cancer cells that result in new and abnormal growth.

Nonhomologous recombination. A mechanism of DNA repair in which two dissimilar broken ends of DNA are ligated using the set of end-binding proteins, polymerase and ligase. This is also called illegitimate recombination.

Oncogene. Those genes that exert a dominant effect in expressing one or more characteristics of malignancy. They are often a result of specific mutations.

8-oxyguanine. The product of an oxygenation reaction from endogenous metabolism or exposure to ionizing radiation that adds an oxygen atom to the 8-position of the guanine base in DNA. This is one of the more common products of oxygenation reactions in DNA and in nucleotide pools.

p53. A protein having a molecular weight of 53 kilodalton. It has a large variety of functions including transcriptional activation, binding to DNA repair proteins and to single stranded DNA, Holliday junctions, and other damaged DNA structures. The gene coding for the p53 protein is conventionally represented in italics as *p53*; a convention which is generally employed in distinguishing proteins from their genes.

Packed tower aeration (PTA). A method for removing volatile contaminants from water by passing a flow of air over a thin film of the water.

PBPK models. Physiologically-based pharmaco-kinetic models: mathematical models that incorporate physiological principles, e.g., blood-flow to tissues, to simulate the movement (kinetic behavior) of contaminants (e.g., radon) in the body.

Phenotype. The visible expression of genetic information contained in the DNA of an organism (i.e., its genotype).

picocurie (pCi). A quantity of radioactivity equivalent to 3.7×10^{-2} decays per second or 2.22 decays per minute. One pCi = 0.037 Bq.

Plug flow reactor (PFR). A treatment unit where the fluid enters the influent end and travels as a discrete packet (plug) to the effluent end of the unit without mixing with packets of fluid ahead or behind it.

Point-of-entry treatment (POE). A process where a contaminant is removed from water just before it is used in an individual household or business, as, opposed to treatment at a central location before the water is distributed to many users.

Potential alpha energy. The total kinetic energy of all the alpha particles emitted by a mixture of radon decay products when all of the atoms in the mixture have completely decayed into ^{210}Pb. Potential alpha energy is measured in Joules (J) or MeV.

Potential alpha energy concentration (PAEC). The concentration of potential alpha energy for radon decay products suspended in a volume air. PAEC is measured in quantities of J m^{-3} or Working Level (1 WL = 2.08×10^{-5} J m^{-3}).

Precision. The uncertainty in a single result from a measurement or procedure that is caused by inherent variability in the processes that are combined to form the result. Precision is a complement to variability.

Preneoplastic. Cells with altered genetic status that are not yet completely malignant.

Probability. A number expressing the chance that a specified event will occur. It can range from 0 (indicating that the event is certain not to occur) to 1 (indicating that the event is certain to occur).

Proliferative cells. Those cells in a tissue that divide by mitosis to become the expanding population that serves as the source for the fully differentiated cells which ultimately carry out the functions of the tissue.

rad. A quantity of absorbed dose equal to 100 erg g^{-1}. This is the traditional unit of dose and 1 rad = 0.01 gray (Gy).

Radiation. Any combination of elementary particles that have sufficient kinetic energy to interact with and transfer energy to material that intercepts their path. If the energy transferred is sufficient to produce ionization in the material, it is classified as ionizing radiation.

Radiation exposure. The total electrical charge of one sign produced in air by photons interacting in volume of air divided by the mass of air in that volume. It has the dimensions of charge divided by mass, Coulomb kg^{-1}. The special unit of exposure is the Roentgen (R).

Radiation weighting factor (w_R). A modifying factor used to obtain equivalent dose from absorbed dose. It depends on the biological effectiveness of the specified radiation but does not depend on the tissue or organ under consideration.

Radioactivity. A quantity of radioactive material based on the *rate* that the atoms or nuclei spontaneously transform and emit radiation. Radioactivity is described in terms of a becquerel (Bq), or curie (Ci).

Radon. The element with atomic number 88. It is an inert gas within the same family of elements as helium, neon, argon, and xenon. Although there are several isotopes of radon, the most common is ^{222}Rn which is part of the decay series that begins with ^{238}U.

Radon decay products. The four short-lived radioactive isotopes in the ^{238}U series immediately following the decay of ^{222}Rn. They are the metals, ^{218}Po, ^{214}Pb, ^{214}Bi, ^{214}Po, and are also called radon daughters or radon progeny.

RCRA. Resource, Conservation and Recovery Act regulates the generation, storage, transportation, treatment and disposal of hazardous substances.

rem. A measure of equivalent dose H, that is obtained when the dose (rad) is multiplied by the radiation weighting factor (w_R). For example, if w_R = 1, then one rem is numerically the same as one rad.

Risk. Conceptually, risk is a measure of the chance that a specified health outcome will occur. Various quantities are used to describe different aspects of risk. For example, the probability of disease is the chance that an individual will develop a specified disease during a selected interval of age (often the entire lifetime). Risks expressed as incidence rates, or mortality rates, are the number of persons who are expected to develop or die from the disease within a selected time interval and population group. This could be expressed as the number of cases or deaths per 100,000 persons per year.

Roentgen. A quantity of radiation exposure equivalent to 2.58×10^{-4} Coulomb kg^{-1}. The roentgen is a traditional unit named on behalf of the German scientist W. K. Roentgen.

Safe Drinking Water Act (SDWA). Legislation that originally became law in 1974 and was amended as recently as 1996. This act and its amendments are designed to provide safe drinking water for consumption by the public.

Senescence. The stage in the life cycle of a cell when division stops and degenerative changes begin to occur. In human cells this happens after approximately 50 cell divisions.

SI units. The International System of Quantities and Units derived by the General Conference of Weights and Measures. Base units are the meter, kilogram, second and coulomb. It is often referred to as the metric system.

Sievert (Sv). A measure of equivalent dose, H, that is obtained when the Dose (Gy) is multiplied by the radiation weighting factor w_R. For example, when w_R = 1, then one Sv is numerically the same as one Gy. The Sievert is an SI unit named on behalf of the Swedish scientist Rolf Sievert.

Signal transduction. The chain of events that starts from initial damage, or from an intercellular signal peptide, and propagates through intervening molecules to eventually cause such events as apoptosis, changes in gene expression or cell cycle delays.

Somatic. The cells of the adult body tissues, in contrast to germ cells which are involved in sperm and egg cell production.

Splanchnic. Pertaining to or affecting the organs in the viscera, especially the intestines, and their associated blood vessels.

Stem cells. The cells in a tissue that have indefinite potential for cell division, and divide relatively slowly. They serve as a source for most of the cells in a tissue and act as a reserve for repopulation following tissue damage.

Stochastic radiation injury. Health effects from ionizing radiation that occur randomly where the *probability* of occurrence is proportional to absorbed dose with no apparent threshold.

Teratogenic. An agent that tends to produce anomalies in developing embryos.

Teratologic. That division of embryology and pathology which deals with abnormal development and congenital anomalies.

Telomere. The specific DNA sequences that form the ends of chromosomes and are replicated by specific enzyme systems, telomerase, which contain a polyribonucleotide sequence that acts as a template for telomere replication.

Threshold. A value of absorbed dose below which the probability of a specific radiation induced health effect is zero.

Time-since-exposure (TSE) model. A risk projection model in which the risk varies with the time after exposure.

Tissue weighting factor (w_T). The ratio of the risk for developing radiation induced cancer in a tissue, T, to the combined risk of developing cancer in any organ following a uniform irradiation of the whole body to the same equivalent dose.

TOC. Total organic carbon: sum of the particulate and dissolved organic matter in a sample such as water.

Transcription. The process by which a gene sequence in DNA is used to synthesize a matching sequence of ribonucleic acid (RNA).

Transfer factor (T). The average increment of the radon concentration in indoor air ($\overline{\Delta C_a}$) divided by the average radon concentration in the water ($\overline{C}_w$).

Translation. The process where a specific RNA sequence is used for the synthesis of a protein in which a code based on 3 bases read together (triplet code) specifies an amino acid position in the protein.

Trihalomethanes (THMs). A class of potentially toxic chemical byproducts such as chloroform, dibromochloromethane, dichlorobromomethane, and bromo-form, that can be formed when water that contains natural organic matter is disinfected by the addition of chlorine.

Triplet repeats. Regions of DNA in which a sequence of three nucleotides is repeated many times.

UCL. Upper confidence limit; the upper bound of a confidence interval.

Uncertainty. A lack of knowledge concerning the truth. This can be quantitative or qualitative.

Variability. The variation of a property or a quantity among members of a population. Such variation is inherent in nature and thus unavoidable. It is often assumed to be random, and can be represented by a frequency distribution.

Volatile organic contaminants (VOCs). Those organic compounds that are considered contaminants in air or water. The compounds usually transfer spontaneously from water to air.

Working level (WL). A quantity of potential alpha energy concentration equivalent to 2.08×10^{-5} J m^{-3}.

Working level month (WLM). An exposure to radon decay products suspended in air that is determined by PAEC multiplied by the residency time at that location. One WLM is equivalent to 1 WL for 170 hours which equals 12.7 J m^{-3} s.

X-ray. A type of electromagnetic radiation, also called a photon, that originates from an energy transition of the atom. These are generally less energetic than gamma rays that are emitted from nuclear transitions.

Conversions between SI units and Traditional Units

Concept	Traditional	SI
Radioactivity	1 Ci	3.7×10^{10} Bq
Radioactivity	1 pCi	0.037 Bq
Concentration	1 pCi L^{-1}	0.037 Bq L^{-1}
Concentration	1 pCi L^{-1}	37 Bq m^{-3}
Absorbed Dose	1 rad	0.01 Gy
Equivalent Dose	1 rem	0.01 Sv
PAEC	1 WL	2.08×10^{-5} J m^{-3}
PAEC Exposure	1 WLM	12.5 J m^{-3} s

Appendixes

A

Behavior of Radon and its Decay Product in the Body

A physiologically based pharmacokinetic (PBPK) model was developed to describe the fate of radon within systemic tissues. A schematic diagram of the model is shown in figure A-1. The model is based on the blood-flow model of Leggett and Williams (1995) (see also Williams and Leggett 1989; Leggett and Williams 1991). The blood of the body is partitioned into a number of compartments, representing various blood pools in the body (the compartment *Large Veins* represents the venous return from the systemic tissues, *Right Heart* and *Left Heart* the content of the heart chambers, *Alveolar* represents the region of gas exchange in the lung, and *Large Arteries* represent the arterial blood flow to the systemic tissues). The gastrointestinal tract is divided into four segments (*St, SI, ULI,* and *LLI*) denote stomach, small intestine, upper large intestine, and lower large intestine, respectively, and *Cont* and *Wall* refer to the content and wall of the segments; for example *St Cont* and *St Wall* denote the content and wall of the stomach. Ingested radon enters the *St Cont* compartment while inhaled radon would enter the *Alveolar* compartment. The walls of the gastrointestinal tract are perfused with arterial blood which, with that from the spleen and pancreas, enters the portal circulation of the liver as shown in figure A-1. Radon dissolved in blood entering the *Alveolar* compartment exchanges with the alveolar air and is exhaled from the body. Although the kinetics of blood circulation are complex, for the most practical purposes it can be viewed as a system of first-order transfer among the different blood pools.

MODEL STRUCTURE

The model, a system of compartments, depicts the manner in which radon is distributed among the tissues of the body and subsequently removed from the

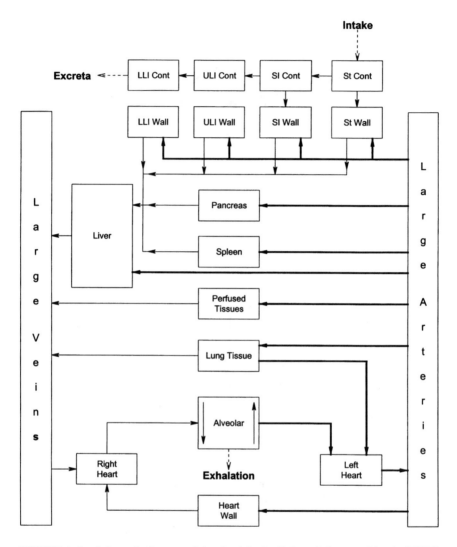

FIGURE A-1   Schematic diagram of the physiologically based pharmacokinetic (PBPK) model developed to describe the fate of radon within systemic tissues.

body. Radon can enter the body either through inhalation (introduced into the *Alveolar* compartment) or by ingestion (introduced into the *St Cont* compartment), only the latter is of interest here. The quantity of radon in organ i, Q_i, is given by a differential equation describing the inflow of radon to the organ and its outflow from the organ. The concentration of radon in organ i, C_i, perfused by blood is assumed to be directly proportional to the radon concentration in the venous blood leaving the organ, $C_{V,i}$. The constant of proportionality between concentration in the organ and the venous blood is referred to as the tissue-blood partition coefficient λ_i defined by

$$\lambda_i = \frac{C_i}{C_{V,i}}.$$

A.1

The equations describing the kinetics of the gas (radon) in perfused tissues are formulated in the manner described by Kety (1951), that is, the rate of change in the quantity of gas in the ith compartment or organ, Q_i, is given by

$$\frac{d}{dt}Q_i = \dot{V}_{CO}F_i(C_A - C_{V,i})$$

A.2

where $\dot{V}_{CO}$ denotes the volumetric flow of arterial blood from the heart (the cardiac output), F_i is the fraction of the cardiac output entering the ith organ, C_A is the concentration of the gas in the arterial blood, and $C_{V,i}$ is the concentration in the venous blood leaving the organ. Rewriting Eqn. A-2 in terms of the activity of radon A_A and A_i in the arterial blood (*Large Arteries* compartment of figure A.1) and in organ i, respectively, and including radioactive decay as a removal mechanism yields

$$\frac{d}{dt}A_i = \frac{\dot{V}_{CO}F_i}{V_A}A_A - \left(\frac{\dot{V}_{CO}F_i}{V_i\lambda_i} + \lambda_R\right)A_i$$

A.3

where V_A is the volume of arterial blood, V_i is the volume of the organ, and λ_R is the decay constant of ^{222}Rn. The volume of an organ is related to its mass M_i as $V_i = M_i / \rho_i$. Eqn. A-3 is applicable to organs perfused by arterial blood and represented collectively, in figure A-1, by the *Perfused Tissues* compartment.

Radon removed from the *Perfused Tissues* compartment enters the compartment *Large Veins* compartment. The rate of change of the radon activity in the *Large Veins* compartment, A_V, is given by

$$\frac{d}{dt}A_V = \sum_i \frac{F_i\dot{V}_{CO}}{V_i\lambda_i}A_i + \frac{F_{lung}\dot{V}_{CO}}{V_{lung}\lambda_{lung}}A_{lung} - \left(\frac{\dot{V}_{CO}}{V_V} + \lambda_R\right)Av$$

A.4

where the summation is over all perfused tissues other than the lung, A_{lung} is the activity of radon within the tissues of the lung (compartment *Lung Tissue*), F_{lung} is the fraction of the cardiac output distributed to lung tissue of which only one-third enters the *Large Vein* compartment, V_{lung} is the volume of lung tissue, and V_V is the volume of venous blood. Note that the fraction of the cardiac output flowing from the liver is the sum of the cardiac output to the segments of the gastrointestinal tract, the spleen, pancreas and the liver itself.

Radon enters the *Right Heart* compartment with the flow of venous blood from the *Large Veins* compartment and the flow of venous blood from the heart (*Heart Wall* compartment). The change in the radon activity in the *Right Heart* compartment, A_{RH}, is given by

$$\frac{d}{dt} A_{RH} = \frac{\dot{V}_{CO}}{V_V} A_V + \frac{F_{HT} \dot{V}_{CO}}{V_{HT} \lambda_{HT}} A_{HT} - \left(\frac{\dot{V}_{CO}}{V_{RH}} + \lambda_R \right) A_{RH} \qquad \text{A.5}$$

where A_{HT} activity of radon present within the tissues of the heart, F_{HT} is the fraction of the cardiac output distributed to heart tissues, V_{HT} is the volume of heart tissues, and V_{RH} is volume of blood within the chambers of the right heart.

Radon may enter into and be removed from the systemic circulation in the gas-exchange regions of the lung. Radon in blood leaving the gas-exchange region of the lung is assumed to be in equilibrium with the alveolar air. The constant of proportionality being the air-blood partition coefficient λ_{air}. The change in the radon activity in the alveolar air, A_V is given by

$$\frac{d}{dt} A_V = \dot{V}_V C_I + \frac{\dot{V}_{CO}}{V_{RH}} A_{RH} - \left(\frac{\dot{V}_V + \dot{V}_{CO} \lambda_{air}}{V_V} + \lambda_R \right) A_V \qquad \text{A.6}$$

where C_I is the concentration of radon in inspired air, $\dot{V}_V$ is the volumetric inhalation rate, V_V is the alveolar volume, and λ_{air} is the air-blood partition coefficient for radon. If the intake is not by inhalation, as in the case of radon in drinking water, then C_I is taken as zero.

Radon enters the *Left Heart* compartment from the *Alveolar* and *Lung Tissue* compartments and departs to *Large Arteries* compartment. The concentration of radon in the blood flowing from the gas-exchange region of the lung, C_P, is in equilibrium with the alveolar air; i.e., $C_P = \lambda_{air} C_V$, where C_v is the radon concentration in alveolar air. The change in the radon activity in the *Left Heart* compartment, A_{LH}, of volume V_{LH} is given by

$$\frac{d}{dt} A_{LH} = \frac{\dot{V}_{CO} \lambda_{air}}{V_P} A_V + \frac{2 \dot{V}_{CO} F_{lung}}{3 V_{lung} \lambda_{lung}} A_{lung} - \left(\frac{\dot{V}_{CO}}{V_{LH}} + \lambda_R \right) A_{LH} \qquad \text{A.7}$$

where λ_{lung} is the lung tissue-blood partition coefficient for radon, and the factor 2/3 represents the fractional of the blood flow from the lung that enters the *Left Heart* compartment.

The gastrointestinal tract model is shown in figure A-1. The equations for the radon activity in the spleen and pancreas follow the equation for perfused organs presented above (equation A.4) and are noted here only because their venous blood enters the liver. Assume that at time zero the activity of radon in the *St Cont* compartment is A_{St}^0, then the changes in the radon activity of the *St Cont* compartment, A_{ST}, is given by

$$\frac{d}{dt} A_{St} = -(k_{St} + k_{Stw} + \lambda_R) A_{St} \; ; A_{St}(0) = A_{St}^0 \qquad \text{A.8}$$

where k_{St} and k_{Stw} are the coefficients for transfer of the gas from the stomach contents to the small intestine contents and to the wall of the stomach, respectively. The value of the coefficient k_{Stw} was derived to correspond to that indicated by the diffusion model discussed in Appendix B. That is, the time integrated concentration of radon in the wall of the stomach was taken to be 30% of that in the content of the stomach. The value of k_{St}, the transfer coefficient from the stomach content to the small intestine contents, for water, was taken to correspond to a half-time of 15 min.

Blood flows through the walls of the segments of the gastrointestinal tract and enters the liver. The change in radon activity within the *St Wall* compartment, A_{StW}, is given by

$$\frac{d}{dt} A_{StW} = k_{St} A_{St} + \frac{F_{StW} \dot{V}_{CO}}{V_A} A_A - \left(\frac{F_{StW} \dot{V}_{CO}}{V_{StW} \lambda_{StW}} + \lambda_R \right) A_{StW} \qquad \text{A.9}$$

where V_{StW} is the volume of the stomach wall.

The equations describing the rate of change in the activity of radon in the contents and walls of the other segments of the gastrointestinal tract have similar form, that is

$$\frac{d}{dt} A_j = k_{j-1} A_{j-1} - (k_{j \, \text{wall}} + \lambda_R) A_j$$

$$\frac{d}{dt} A_{j \, \text{wall}} = k_j A_j + \frac{F_{j \, \text{wall}} \dot{V}_{CO}}{V_A} A_A - \left(\frac{F_{j \, \text{wall}} \dot{V}_{CO}}{V_{j \, \text{wall}} \lambda_{j \, \text{wall}}} + \lambda_R \right) A_{j \, \text{wall}}$$

where $j = SI$, *ULI*, and *LLL* denote the regions of the tract.

PARAMETER VALUES

Adult Values

The first-order transfer coefficients describing the movement of radon within the blood are, as indicated above, dependent on the cardiac output $\dot{V}_{CO}$, the distribution of the cardiac output F_i, and the tissue-to-blood partition coefficient λ_i. The reference values for the total blood volume and cardiac output in an adult male are 5.3 L and 6.5 L min⁻¹, respectively (Leggett and Williams 1995). The large arteries and veins in Fig A-1 contain 6 and 18% of the blood volume of the body, respectively. The distribution of the cardiac output (Leggett and Williams 1995) and the tissue-to-blood partition coefficients (Nussbaum 1957) for the various organs are given in table A-1. The masses and densities of the organs in the adult male are given in table A-2. As an example, table A-1 indicates that 0.3% of the cardiac output is directed to the adrenals which have a radon partition coefficient of 0.7, thus the coefficient transferring radon from the large arteries to

the adrenals, the terms $\dfrac{\dot{V}_{co}F_i}{V_i}$ of Eqn. A-4, has a value of

$$\frac{6.5 \text{ L min}^{-1} \times 0.003 \times 1.02 \text{ g cm}^{-3} \times 10^3 \text{cm}^3 \text{ L}^{-1}}{0.014 \text{ kg} \times 10^3 \text{g kg}^{-1} \times 0.7} = 2.03 \text{ min}^{-1}$$

where 0.06×5.3 L is the volume of blood in the large arteries. The removal coefficient from the adrenals to the large veins, the term $\dfrac{\dot{V}_{co}F_i}{V_i\lambda_i}$ or $\dfrac{\dot{V}_{co}F_i\rho_i}{M_i\lambda_i}$ of Eqn. A-4, is

$$\frac{6.5 \text{ L min}^{-1} \times 0.003}{0.06 \times 5.3 \text{ L}} = 0.612 \text{ min}^{-1}$$

where all numerical values are from tables A-1 and A-2.

Radon is considerably more soluble in adipose tissue than other tissues of the body as reflect in the high adipose-to-blood partition coefficient listed in table A-1. The transfer of radon from the large arteries to adipose tissue, the terms $\dfrac{\dot{V}_{co}F_i}{V_i}$ of Eqn. A-4, has a value of

$$\frac{6.5 \text{ L min}^{-1} \times 0.55}{0.06 \times 5.3 \text{ L}} = 1.022 \text{ min}^{-1} \qquad\qquad \text{A.9a}$$

where is the volume of blood in the large arteries. The removal coefficient from the adipose tissue to the large veins, the term $\dfrac{\dot{V}_{co}F_i}{V_i\lambda_i}$ or $\dfrac{\dot{V}_{co}F_i\rho_i}{M_i\lambda_i}$ of Eqn. A-4, is

$$\frac{6.5 \text{ L min}^{-1} \times 0.05 \times 0.92 \text{ g cm}^{-3} \times 10^3 \text{cm}^3 \text{ L}^{-1}}{12.5 \text{ kg} \times 10^3 \text{g kg}^{-1} \times 11.2} = 2.14 \times 10^{-3} \text{ min}^{-1} \quad \text{A.10}$$

where all numerical values are from tables A-1 and A-2. The biological removal from adipose tissue corresponds to a half-time, in the absence of additional input, of about 5.4 hours.

TABLE A-1   Reference Regional Blood Flows (% of Cardiac Output) and Radon Tissue-to-Blood Partition Coefficients in PBPK Model.

Compartment	Flow (%)	λ_i	Compartment	Flow (%)	λ_i
Stomach Wall	1.0	0.7	Kidneys	19.0	0.66
Small Intestine Wall	10.0	0.7	Muscle	17.0	0.36
Upper Large Intestine Wall	2.0	0.7	Red Marrow	3.0	8.2
Lower Large Intestine Wall	2.0	0.7	Yellow Marrow	3.0	8.2
Pancreas	1.0	0.4	Trabecular Bone	0.9	0.36
Spleen	3.0	0.7	Cortical Bone	0.6	0.36
Adrenals	0.3	0.7	Adipose Tissue	5.0	11.2
Brain	12.0	0.7	Skin	5.0	0.36
Heart Wall	4.0	0.5	Thyroid	1.5	0.7
Liver	6.5	0.7	Testes	0.05	0.43
Lung Tissue	2.5	0.7	Other	3.2	0.7

TABLE A-2   Mass and Density of Organs in the Adult Male

Compartment	Mass (kg)	ρ_i	Compartment	Mass (kg)	ρ_i
Stomach Wall	0.15	1.05	Kidneys	0.31	1.05
Small Intestine Wall	0.64	1.04	Muscle	28.0	1.04
Upper Large Intestine Wall	0.21	1.04	Red Marrow	1.5	1.03
Lower Large Intestine Wall	0.16	1.04	Yellow Marrow	1.5	0.98
Pancreas	0.10	1.05	Trabecular Bone	1.0	1.92
Spleen	0.18	1.05	Cortical Bone	4.0	1.99
Adrenals	0.014	1.02	Adipose Tissue	12.5	0.92
Brain	1.4	1.03	Skin	2.6	1.05
Heart Wall	0.33	1.03	Thyroid	0.02	1.05
Liver	1.8	1.04	Testes	0.035	1.04
Lung Tissue	0.47	1.05	Other	3.2	1.04

The response of the model was shown graphically in chapter 4 and compared to experimental observations.

Other Ages

Considerably less information is available regarding cardiac output and blood volumes in the non-adult. Age-dependence in the model was introduced by assuming the blood flow to an organ was proportional to the mass of the organ; the constant of proportionality being derived from the adult values. The age-dependent masses were taken from ICRP Publication 56 (1989). The cardiac output was taken to be 0.6, 1.2, 3.7, 5.0, 6.2, and 6.5 L/min in the newborn, 1-, 5-, 10-, 15 year-old, and adult, respectively (Williams 1993). The volume of blood in the body was taken as 0.27, 0.5, 1.4, 2.4, 4.5, and 5.3 L in the newborn, 1-, 5-, 10-, 15 year-old, and adult, respectively (Williams 1993). In the absence of information on the fraction of the blood volume associated with the large arteries and veins of the body at various ages the adult fractions were assumed; namely 6% and 18% in the arteries and veins, respectively.

BIOKINETICS OF SHORT-LIVED RADON DECAY PRODUCTS

In its recent reports, the International Commission on Radiological Protection (ICRP) have assessed the component of the dose associated with decay products from within the body following the intake of a radionuclide based on the biokinetic behavior of the specific decay product; so-called *independent kinetics*. Details regarding this implementation are discussed in Annex C of Publication 71 (ICRP 1995). In the publication are also set forth the description of the biokinetic of lead, bismuth, polonium, and astatine as members of the uranium decay series. These data have been applied in the dosimetric analysis in this report.

B

A Model for Diffusion of Radon Through the Stomach Wall

The primary functions of the stomach in the gastrointestinal system are to dissolve ingested foods and pass the contents to the small intestine, where nutrients are transferred to the bloodstream. Those functions are facilitated by the production and secretion of acid and enzymes that eventually dissolve the contents. The wall of the stomach contains the cells that generate the acid. The epithelial surface of the stomach wall is coated with a mucous layer that serves as a barrier between the acidic lumen and the tissues surrounding the stomach. Adjacent to the mucous layer is a region of tissue consisting of crypts that contain proliferating stem or progenitor cells that eventually reach the surface to perform the necessary functions. These cells are radiosensitive and are believed to be responsible for initiation and promotion into stomach cancer.

The failure of most materials in the lumen of the stomach to penetrate into the wall in effect serves as a protective measure for stem cells, in that alpha particles originating in the stomach contents cannot reach the stem cells. However, alcohol, aspirin, and inert gases, such as carbon dioxide and the noble gases, are known to penetrate into and pass through the wall of the stomach and to enter the blood stream. The mechanism of this process is assumed to be molecular diffusion, but blood flow can influence clearance through the stomach wall. It is thought that some of the radon ingested with water will also diffuse through the contents to the wall and then through the stomach wall; this presents an opportunity for alpha particles from radon decay to deposit energy at the location of radiosensitive cells.

The objective of this modeling exercise was to estimate the transport of radon through the stomach wall by diffusion. It was not designed to give a complete description of the very complicated geometry and dynamics of the stomach. However, it was created to provide indications of the concentration and duration of radon in regions that may contain radiosensitive cells. These efforts were necessary because no other theoretical or experimental information is available. By varying the parameters, it was possible to obtain a range of results and to identify extreme values that could serve as bounds for radon concentrations in the stomach wall.

For simplicity, the lumen of the stomach was considered to be a sphere. The sphere was surrounded by concentric spherical shells representing the mucous layer and the wall. The stomach was filled with water containing a unit concentration of radon at time $t = 0$. The radon concentration in the mucous and wall was zero at $t = 0$. The radon concentration at the outer surface of the wall was considered to be zero at all times because of the removal of radon by blood flowing through the stomach wall.

The time-dependent equation describing the concentration of radon can be derived from Fick's law:

$$\frac{dC}{dt} = D\nabla^2 C - \lambda C \qquad\qquad \text{B.1}$$

where

C = the concentration of radon,
D = the effective diffusion coefficient,
∇^2 = the Laplacian operator, and
λ = the radioactive-decay constant associated with ^{222}Rn.

Since the intervals associated with events in the stomach are generally much less than the half-life of ^{222}Rn, radioactive decay is neglected. Using spherical symmetry, the Laplacian operator can be expressed in terms of only the radius, r:

$$\nabla^2 = \frac{\partial^2}{\partial r^2} + \frac{2}{r}\frac{\partial}{\partial r} \qquad\qquad \text{B.2}$$

One procedure for solving the equation is to separate C into the product of two components, one in radius only, $R(r)$, and another in time only, $T(t)$ (Andrews 1986). Thus,

$$C(r,t) = R(r)\cdot T(t) \qquad\qquad \text{B.3}$$

and

$$\frac{1}{T}\frac{dT}{dt} = \frac{D\nabla^2 R}{R} = K \qquad\qquad \text{B.4}$$

where K is referred to as a separation constant. Initial and boundary conditions are expressed as

$$C(r,t) = f(r) \qquad \text{when } t = 0 \qquad\qquad \text{B.5}$$

and

$$C(r,t) = h(t) \qquad \text{when } r = a. \qquad\qquad \text{B.6}$$

Substituting $U = r \cdot R$ into the spatial part of the equation transforms the spherical Laplacian operator into a simple second-order differential equation:

$$\frac{d^2 U}{dr^2} + \frac{K}{D}\frac{dU}{dr} = 0. \qquad\qquad \text{B.7}$$

That is identical with the equation for one-dimensional diffusion through a slab, provided that the initial and boundary conditions reflected in U are

$$U(r,t) = r \cdot f(r) \qquad \text{when } t = 0 \qquad\qquad \text{B.8}$$

and

$$U(r,t) = a \cdot h(t) \qquad \text{when } r = a. \qquad\qquad \text{B.9}$$

If a solution for U can be found, then the corresponding solution for the sphere is:

$$C(r,t) = \frac{U}{r} \cdot T. \qquad\qquad \text{B.10}$$

The first example is that for a homogeneous sphere and shell. The shell has no radon at time $t = 0$. The purpose of this calculation is to show how fast radon would escape from an undisturbed sphere and through a shell by diffusion only. The dimensions and initial and boundary conditions are as follows:

r_s(sphere) = 3.6 cm   volume = 200 mL,
r_{ss}(sphere+shell) = 3.7 cm Δr = 1 mm, and
$C(r,t) = C_0 = 1$ when $t \geq 0$.

A solution that satisfies those conditions is:

$$C(r,t) = \frac{2C_0}{r}\sum_{n=1}^{\infty}\left\{\frac{r_{ss}}{n^2\pi^2}\sin\frac{n\pi r_s}{r_{ss}} - \frac{r_s}{n\pi}\cos\frac{n\pi r_s}{r_{ss}}\right\}\sin\frac{n\pi r}{r_{ss}}\exp\left(\frac{-Dn^2\pi^2 t}{r_{ss}^2}\right). \quad \text{B.11}$$

Figure B.1 shows a graph of $C(r)/C_0$ as a function of radius from the center of the sphere with a diffusion coefficient for radon in water of $D = 1 \times 10^{-5}$ cm^2 s^{-1} (Tanner 1964). The radon concentration decreases near the boundary of the sphere but is almost unchanged at a radius less than 3 cm. Even after 1 h, there is considerable radon remaining in the sphere. This illustrates that diffusion alone is not sufficient to transfer all the available radon through the stomach wall in the amount of time corresponding to normal residency in the lumen.

The model was then revised to take into account removal of radon from the lumen by means other than diffusion. The contents of the stomach are considered

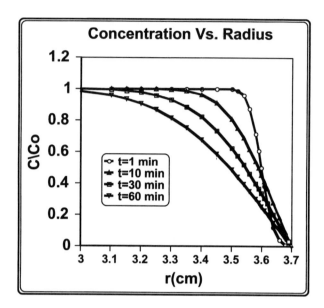

FIGURE B.1   Concentration of radon, C/C_0, vs. radius, r, within a sphere of water with radius $r_s = 3.6$ cm. The interior of the sphere is surrounded by a spherical shell of water with a thickness 0.1 cm and a radon concentration $C(r) = 0$ at $t = 0$. The diffusion coefficient for radon in water is $D = 1 \times 10^{-5}$ cm^2 s^{-1}.

to be well mixed, but the concentration is decreasing exponentially with a half-time of 20 min. For simplicity, the volume of the lumen remains constant such that any material leaving the stomach is replaced with water that dilutes the radon.

The basic differential equations are unchanged. However, the interior surface of the stomach wall is driven by the function $h(t)$ that is controlled by the concentration in the lumen. A solution can be obtained with Duhamel's theorem where the concentration in the wall is the convolution of the time derivative of a solution after a unit step function at $t = 0$ (Carslaw and Jaeger 1959).

The dimensions corresponding to the boundary conditions following an intake of 250 mL of water are as follows (see fig. B.2):

r_s(lumen) = 3.908 cm   volume = 250 mL
r_w(lumen+wall) = 3.938 cm   wall thickness = 0.03 cm (300 μm)
$r_{stem\ cells}$ = 3.928,   depth in wall = 0.02 cm (200 μm)
$C(r,t) = C_0$ when $t = 0$ and $r \leq r_s$
$C(r,t) = C_0\, e^{-\alpha t}$ when $t > 0$ and $r \leq r_s$
                                    $\alpha = 0.000578$ s^{-1}
$C(r,t) = 0$ when t ≥ 0 and $r = r_w$

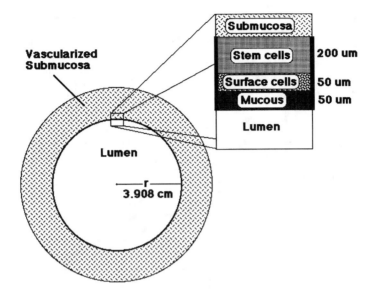

FIGURE B.2   Diagram showing the geometry used in the model for diffusion of radon through the stomach wall.

The wall consists of mucous (50 μm), surface cells (50 μm) and tissue with crypts that contain the radiosensitive stem cells (200 μm) (E. Robbins 1998, personal communication). The stem cells are centered at a depth of 200 μm below the surface. It is assumed that blood flow removes all radon at a depth of 300 μm.

A solution for that situation is

$$C(r,t) = \frac{C_0 r_s}{r} \sum_{n=1}^{\infty} \frac{\dfrac{2Dn\pi}{(r_w - r_s)^2}}{\dfrac{Dn^2\pi^2}{(r_w - r_s)^2} - \alpha} \sin\left(\frac{n\pi(r - r_s)}{r_w - r_s}\right) \left\{ \exp(-\alpha t) - \exp\left(\frac{-Dn^2\pi^2 t}{(r_w - r_s)^2}\right) \right\}. \quad \text{B.12}$$

The results depend on the selection of a diffusion coefficient for radon in the mucous and wall. The effective diffusion coefficient includes a retardation factor that accounts for absorption in the medium. There are no published values in the literature for the effective diffusion coefficient of radon in tissue. For this report, we have adopted a value obtained for xenon and assume a nominal value of $D = 5 \times 10^{-6}$ cm^2 s^{-1} for both the mucous and the stomach wall. The exact location of the radio-sensitive cells is also unknown. For this calculation, the absorbed dose is estimated at a depth of 200 μm which is 150 μm below the mucous.

Figure B.3 shows the results from the model with the concentration $C(r)/C_0$ plotted at various times after intake. It can be seen that the concentration in the

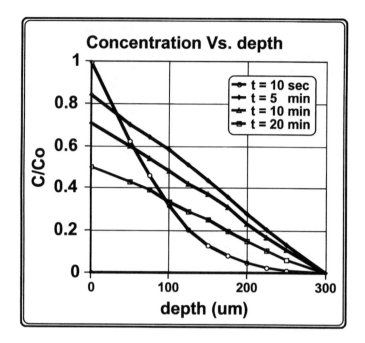

FIGURE B.3   Concentration of radon, C/C_0, as function of depth in the stomach wall. The radius of stomach lumen is 3.908 cm, corresponding to a volume of 250 mL. The concentration in the lumen is considered to be uniform and is assumed to decrease exponentially with a half-time of 20 min. The depth of the mucous layer is 50 μm, and that of the stem cell population is 200 μm.

wall becomes linear with distance from the surface when t is greater than 5 min. Figure B.4 shows the results of the model with the concentration $C(t)/C_0$ plotted as a function of time at various distances (in micrometers) from the inner surface of the wall.

The flow rate of radon through the stomach wall can be obtained from the following relationship:

$$\phi(r,t) \;=\; -D\nabla C(r,t) \qquad\qquad\qquad \text{B.13}$$

where $\phi(r,t)$ is the flow rate of radon per unit area, often referred to as the flux density. Because radioactive decay has been neglected, one would expect the time-integrated flux to yield the total radon flowing through the wall and this should be independent of r:

$$\Phi(r) \cdot A = \int_0^\infty \phi(r,t)dt \cdot 4\pi r^2 = \text{constant.} \qquad\qquad \text{B.14}$$

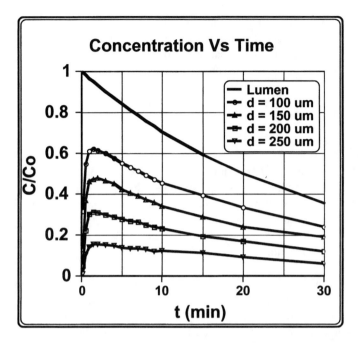

FIGURE B.4   Concentration C/C_o in the stomach wall as a function of time after intake. The radius of the stomach lumen is 3.908 cm, corresponding to a volume of 250 mL. The concentration in the lumen is considered to be uniform and is assumed to decrease expo-nentially with a half-time of 20 min. The depth of the mucous layer is 50 μm and that of the stem cell population is 200 μm.

For the conditions described here, the fraction of radon released to the blood stream due to diffusion through the stomach wall is 20%.

The absorbed dose to the cells at a depth, r, is related to the time-integrated radon concentration at this same depth. The integrated concentration, relative to that in the lumen is

$$\frac{\int\limits_{0}^{\infty} C(r,t)dt}{\int\limits_{0}^{\infty} C_0 e^{-\alpha t}dt} = \alpha \int\limits_{0}^{\infty} \frac{C(r,t)}{C_0} dt \qquad \text{B.15}$$

For conditions described here, the time-integrated concentration at the location of the stem cells (depth = 200 μm) was approximately 30% of the integrated concentration in the lumen. As seen from figure B.3, the time-integrated concentration varies linearly with depth through the stomach wall. The results do not change significantly when the diffusion coefficient is varied from 10^{-5} to 10^{-7} cm^2 s^{-1}. The number of radon atoms that decay in the vicinity of the stem cells can be obtained by

$$N(\alpha) = \int_0^\infty \frac{C(r_s, t)dV_s}{C_0 V_{\text{Lumen}}} dt, \qquad \text{B.16}$$

where

$N(\alpha)$ = the number of nuclear transformations of ^{222}Rn occurring in the volume element dV per ingested Bq,

dV_s = the volume of a spherical shell surrounding the assumed location of stem cells in the stomach wall $(4\pi r^2 \, dr)$,

r = 3.928 cm, at a depth of 200 μm, and

dr = 100 μm.

The result of this integration yields four nuclear transformations per becquerel of ^{222}Rn after ingestion of 250 mL of water. That indicates that energy deposition by alpha particles in the vicinity of the radiosensitive cells will certainly not be uniform. Absorbed dose obtained by averaging energy deposition over the volume of interest for this situation should be interpreted with caution.

The model presented here assumes that there is no capillary involvement in the first 250 μm of tissue below the mucous layer in the stomach wall. If such capillaries were present in the region between the surface cells and the crypts containing the stem cells, the capillary blood flow would reduce radon penetration into the wall.

It must be emphasized that this is a very naive representation of the actual conditions in the stomach after an intake of water containing ^{222}Rn. However, these simplifications can increase our understanding of the processes associated with the ingestion of radon by illustrating how assumptions about diffusion could influence internal dosimetry. These results can also provide a basis for the development of more-representative models.

C

Water-Mitigation Techniques

AERATION

A wide array of aeration methods can be used to remove radon from drinking water (table C.1). Some are already being used in the United States to treat radon in municipal drinking water (table 8.1); others have been used only in point-of-entry applications or are still being developed.

Packed-Tower Aeration

The most common technology currently used for treatment of large flows of water with high radon concentrations is packed-tower aeration (PTA). PTA is efficient because it has a high surface area available where mass transfer can occur. Usually, raw water is sprayed into the top of the tower (3-9 m high) and trickles down over plastic packing (for example, rings and saddles) that has a high ratio of surface area to volume. Simultaneously, a flow of air is pumped through the packing. Typically, this is a countercurrent flow of air, which enhances radon removal. The treated water is collected in a reservoir below the tower and pumped to a pressurized storage tank or directly into the distribution system. Air containing the radon is released from the top of the tower. One variant on PTA is cross-current technology in which the air flow is perpendicular to the water trickling down, so less energy is required to supply a given amount of air to the system.

Diffused-Bubble Aeration

Diffuser systems inject air (usually as bubbles) into water. The radon moves from the water to the bubbles as they rise through the liquid. Smaller bubbles,

TABLE C.1 Aeration Technologies Used for Removing Radon from Water

Common Name	Other Name
Packed tower aeration (PTA)	—
Diffused bubble aeration	—
Spray aeration	—
Tray aeration	Slat-tray aeration
Jet aeration	Venturi or ejector aeration
Shallow-tray aeration	Sieve tray aeration
Cross-current packed-tower aeration	—
Cascade aeration	—
Pressure aeration	Aeration in hydrophor degasification

although more difficult and expensive to produce, provide a greater surface area per unit volume over which mass transfer can occur. In the most common systems, the water passes through a series of tanks (0.5-6 m deep), simulating a plug-flow reactor. The radon-contaminated air leaves the water when the bubbles reach the surface and is vented out of the unit. Diffused aeration systems cannot match the surface area available for radon transfer from water to air in PTA, but they can be easily retrofitted into basins and made as compact package units to treat small to medium flows. Shallow-tray (20-30 cm deep) aeration is a variant of this technology in which a thin layer of water passes across a series of plates perforated with holes. Air coming up through the holes causes the water to froth, and mass transfer occurs.

Spray Aeration

In spray aeration, water is formed into droplets (with a high ratio of surface area to volume) when it is forced through a nozzle. The droplets are sprayed upward, downward, or at an inclined angle into a large volume of air that is often flowing in a countercurrent direction. The simplicity of spray-aeration systems means that they can easily be retrofitted onto the inlet of an existing atmospheric storage tank to enhance radon removal from water. Their radon-removal efficiency is mainly a function of the size of the water droplets and the ratio of air to water (A:W ratio).

Tray Aeration

Tray-aeration systems are similar to countercurrent PTA except that the tower contains a series of slats (for example, made of redwood) or trays with perforated bottoms (for example, made of wire mesh). In some cases, a solid medium (such as, stone, ceramic spheres) is placed in the trays to promote transfer of radon to the air (Drago 1998). Water entering the top of the aerator is distributed over the trays or slats. Natural, forced, or induced draft causes air to flow past the thin film of water formed.

Cascade Aeration

Cascade aeration, a very simple technology, involves construction of a series of steps over which water tumbles as in a waterfall. The system requires a hydraulic head to operate, but little else except a method of ventilation to remove the radon-contaminated air from the unit.

Jet Aeration

Use of jet-aeration systems is favored in Europe, where they can be retrofitted into existing small storage-tank systems. The water is pumped through a venturi-like device (such as a jet, eductor, or ejector) that aspirates air into the water. The radon-contaminated air is released, and the treated water falls into an atmospheric storage tank. The water must be recirculated through the system several times before high removal efficiencies (>75%) are obtained. A venturi system tested on two US water supplies achieved radon-removal efficiencies of 78-95%.

Pressure Aeration

In pressure aeration, air is injected into a pressurized chamber [for example, tank (hydrophor), pipe]. The gas is released when the water is allowed to come to atmospheric pressure. Although this technology uses lower A:W ratios (1:1) compared to other aeration methods (10:1 to > 100:1), it might be applicable only to special situations because the energy required to inject the air can be very high.

GRANULAR ACTIVATED CARBON

Granular activated carbon (GAC) is made by subjecting materials such as bone, wood, or coconut shells to high heat and pressure. These processes increase the surface area of the material and activate it, improving its ability to adsorb substances, including organic chemicals, and dissolved gases. GAC has a finite number of sites where it can adsorb a specific substance. Hence, it normally becomes saturated with the contaminant that it is removing from water over the course of days to months and must be replaced or regenerated (for example, by steam-cleaning when used to remove volatile organic compounds) to sustain an adequate level of treatment.

Adsorption of radon from water does not follow the typical saturation model observed for many contaminants, but instead can be modeled with a steady-state first-order equation first proposed by Lowry and Brandow (1985). The relationship can be used to calculate important design variables, such as the empty-bed contact time (EBCT) and the volume of GAC needed to reach a desired effluent quality.

$$EBCT(h) = \frac{\ln\left(\frac{C_t}{C_o}\right)}{K_{ss}} = \frac{V}{Q}$$ C.1

where

C_t = effluent radon concentration (kBq m^{-3}),
C_o = influent radon concentration (kBq m^{-3}),
K_{ss} = the adsorption/decay constant specific for the GAC and the water treated (h^{-1}),
V = volume of GAC needed (m^3), and
Q = flow of water treated (m^3 h^{-1}).

It has been suggested that saturation does not occur because the radon decays, allowing a new atom of the gas to be adsorbed to the same GAC site (Lowry and Brandow 1985). Others have suggested that very long run times occur before saturation because of the very low mass of radon being adsorbed (Kinner and others 1993; Cornwell and others 1999).

When used for small flows, the carbon is usually placed in a closed vessel (constructed of, for example, fiberglass or carbon steel), and the water is forced through the bed, using the pressure exerted by the well pump. Therefore, re-pumping is not required, because there is no break to atmospheric pressure. In large municipal facilities, operated at atmospheric pressure, the hydraulic head from the water above the GAC causes the water to flow past the GAC. In either system, head-loss problems resulting from accumulation of turbidity-causing substances or precipitates can be alleviated by backwashing. The effect of backwashing on radon removal is not clear; some studies have shown a decline in efficiency after backwashing (Lowry and Brandow 1985) (Lowry and others 1990), and others have not (Kinner and others 1990; Cornwell and others 1999). Lowry and others (1990) have observed desorption of radon during and immediately after backwashing, but the radon progeny remain sorbed.

When the efficiency of a GAC unit in removing radon from water begins to decline (Lowry and others 1991; Kinner and others 1989; Kinner and others 1990; Kinner and others 1993; Cornwell and others 1999), the GAC is usually not regenerated, although it is technically possible to remove accumulated ^{210}Pb by using an acid pumped through the bed or by thermal desorption (Lowry and others 1990). Instead, it is usually easier to dispose of the old carbon before it accumulates a significant amount of radioactivity and add new GAC.

Vacuum Deaeration and Hollow-Fiber Membrane Technology

Vacuum deaeration (VD) exploits the high Henry's Law constant of radon by spraying the raw water into an enclosed tower that contains a packing material. A vacuum is applied to the top of the unit by an eductor or pump. The vapor is redissolved with an eductor into a continuously recirculating stream of water that passes through the GAC bed. Noncondensable gases (such as O_2, N_2, and CO_2) are released from the sidestream via a constant-head tank and an oil trap. The efficiency of radon removal from the water is strongly linked to the strength of the vacuum. At high vacuums (under 0.1 atm absolute pressure), removals in the 70% range have been observed. Two difficulties with the process are the low efficiency of transfer of radon into the stream of recirculating water and the desorption of radon from the GAC (only 20-30% net radon removal observed). Implementation of this complex technology, which would be applicable only to medium and large flows, must await further testing.

The hollow-fiber membrane (HFM) technology is equally complex and differs from the VD process only in using a column of membranes, instead of a tower with packing, to remove radon from water initially. The raw water passes along one side of a series of microporous membranes. A stream of air induced by a vacuum passes along the other side. The radon and other dissolved gases are transferred to the air under vacuum. Again, the efficiency of transfer is a function of the strength of the vacuum applied. With a bench-scale system, a radon removal efficiency of 40-56% was obtained (Drago 1997). The problems with dissolving the radon in the sidestream and the removal efficiency of the GAC (40-80%) observed in evaluations of the HFM system are similar to those for the VD system. Applicability of HFM must be evaluated on pilot- and full-scales before it could be considered a best available technology for radon removal for medium and large communities, which need to remove radon from water to avoid discharging it to the atmosphere via the off-gas.

D

Risks Associated with Disinfection By-products Formed by Water Chlorination Related to Trihalomethanes (THMs)

1) Ingestion of Household Water with THMs

General Assumptions

Drinking-water intake: 0.6 L d^{-1}
Body weight: 70 kg
Person drinks water at home 7 d wk^{-1}, 52 wk y^{-1}
Exposure duration: 70 y
Averaging time for carcinogenic effects: 70 y (25,550 d)

Unit Dose Factor for Cancer Risk (lifetime average):

$$\frac{0.6 \text{L/d}}{70 \text{ kg}} \times C_W = 0.0086 \ \mu g \ / \ (kg - d) \text{ per } \mu g \ / \ L \text{ in the water supply}$$

2) Inhalation Intake from Household Water with THMs

Assumptions

Inhalation rate of moderately active people: 0.77 m^3 h^{-1}
Duration of time in house during a day: 17 h

Exposure duration: 70 y
Averaging time for carcinogenic effects: 70 y (25,550 d)
Average transfer coefficient = 0.1 µg / m³ (air)/µg/L (water)

Unit Dose Factor for Cancer Risk (lifetime average):

$$\frac{0.77 m^3/h \times 17 h/d}{70 kg} \times 0.1 \frac{\mu g/m^3 (air)}{\mu g/L(water)} \times C_w = 0.019 \; \mu g/(kg-d) \; per \; \mu g/L \; in \; the$$

water supply

3) Dermal Uptake from Bathing at Home

Assumptions

Body weight: 70 kg
Area of body exposed during bathing: about 20,000 cm² (in the range of
central to upper estimate from EPA guidance)
Effective permeability of THMs through skin for a 10-min exposure:
0.05 cm h^{-1}, based on value for chloroform estimated by McKone
(1993) from EPA and other models
Bath or shower for 0.17 h (10 min) once per day, 365/day
Exposure duration: 70 y
Averaging time for carcinogenic effects: 70 y (25,550 d)

Unit Dose Factor for Cancer Risk (lifetime average) =

$$\frac{0.05 cm/h \times 0.17 h \times 20,000 cm^2}{70 kg} \times \frac{1 \; event}{day} \times \frac{1L}{1000 cm^3} \times C_w$$
$$= 0.0024 \mu g/(kg-d) \; per \; \mu g/L \; in \; the \; water \; supply$$

In the tables below, Risk = C_w × UDF × Potency.
The total risk for each chemical is the sum of the risks across the three
exposure routes.

Summary Table of Risk Estimates for THMs in Surface Water

| | Unit Dose Factor | Compound | | | | |
		Chloro-form	Bromo-dichloro-methane	Chloro-dibromo-methane	Bromo-form	Total Risk
Water Concentration[a] $\mu g\ L^{-1}$		90	12	5	2.1	
		Potencies by Route				
Ingestion	0.0086	0.031	0.13	0.094	0.008	
Inhalation	0.019	0.019	0.13	0.094	0.004	
Dermal uptake	0.0024	0.019	0.13	0.094	0.004	
Total risk by chemical		6.1×10^{-5}	4.7×10^{-5}	1.4×10^{-5}	3.2×10^{-7}	**1.2×10^{-4}**

[a] Data from Brass and others (1981).

Summary Table of Risk Estimates for THMs in Ground Water

| | Unit Dose Factor | Compound | | | | |
		Chloro-form	Bromo-dichloro-methane	Chloro-dibromo-methane	Bromo-form	Total Risk
Water Concentration[a] $\mu g\ L^{-1}$		8.9	5.8	6.6	11	
		Potencies by Route				
Ingestion	0.0086	0.031	0.13	0.094	0.008	
Inhalation	0.019	0.019	0.13	0.094	0.004	
Dermal uptake	0.0024	0.019	0.13	0.094	0.004	
Total risk by chemical		6.0×10^{-6}	2.3×10^{-5}	1.9×10^{-5}	1.7×10^{-6}	**4.9×10^{-5}**

[a] Data from Brass and others (1981).

E

Gamma Radiation Dose From Granular-Activated Carbon (GAC) Water Treatment Units

The committee has used an extended-radiation-source program to estimate the annual gamma radiation dose from water treatment units containing granular activated carbon (GAC) for water-treatment-plant workers or for situations where point-of-entry (POE) units are used. The extended source accounts for the distribution of the radioactivity throughout the bed instead of assuming that the radioactivity is located at a single point. This program, MICRO-SHIELD (Grove Engineering Company, Rockville, MD), was used to compute gamma ray exposure rates at locations outside a volume containing radioactivity with a specified distribution of radioisotopes. The program accounts for both the source-receptor geometry and the attenuation of the gamma rays by the materials within the tank (such as water and carbon) and the tank wall. The program also contains several options to correct for the second-order effects of a thick source and attenuation; in this case, the Taylor approximation for build-up was used.

For the present calculations, the committee used a vertical, cylindrical tank with a 1-cm-thick iron wall. The radius and length of the tank were chosen based on standard engineering designs for the POE or water treatment plant flow rates. These water flow rates and tank dimensions are shown in table E.1. Flow rates for POE systems are typically 1 m^3/d, while flow rates for water treatment plants using GAC range from 11 to 981 m^3/d. Two entries are shown for the highest flow rate. The first represents a tank designed for a pressurized treatment system while the second is for a water treatment tank operating at atmospheric pressure with flow driven by gravity. All of the calculations assumed an input radon-in-water concentration of 185 kBq/m^3 and an output concentration of 25 kBq/m^3.

TABLE E.1   Estimated Equivalent Dose Rates and Dose at Water-Treatment Plants or in Point-Of-Entry Applications Using GAC to Remove Radon

Flow (m^3 d^{-1})	Tank Radius (cm)	Tank Height (cm)	Case 1: Equivalent Dose at 1 m (μSv/h)	Case 1: Time to Acquire 1mSv (h)	Case 2: Equivalent Dose^a at 1 m (μSv/h)	Case 3: Equivalent Dose at 1 m (μSv/h)
1 (POE)	12.7	54.5	0.124	8064	0.137	0.068
11	22.8	185	0.666	1488	0.725	0.387
981 (P)^b	91.5	520	5.12	192	7.02	7.01
981 (G)^b	152.5	186	4.69	216	6.35	4.01

^{a}The radiation weighting factor of 1.0 was used for these gamma rays.
^{b} P is pressure-driven while G is gravity fed.

The absorbed dose calculation was performed for a point 1 m from the outside of the tank wall and at the mid-point of the tank height. By specifying that the radiation source is Rn-222, the MICRO-SHIELD program computes the source strength for the various radon decay products, assuming that they are in equilibrium with the radon in the carbon bed. Sufficient time was allowed to elapse to permit the radon decay products to reach equilibrium with the radon. This calculation also ignores the very small contribution to the radiation field made by the longer-lived Pb-210 (22.3 years) and its subsequent decay products.

Doses and dose rates were calculated for several different source assumptions. The first case used a 50-50 (by mass) mixture of carbon and water (with a carbon density of 0.42 g/cm^3) and assumed that the radioactivity was uniformly distributed throughout the cylinder. The results of these calculations are shown as Case 1 in table E.1. The equivalent dose rate is shown for the point at 1 m from the tank wall and at half the height of the tank. In addition, the total time to acquire an annual dose of 1 mSv is shown.

The second set of calculations were done using an "idealized" mixture of water and carbon to give a mixture density of 1.2 g/cm^3 (based on the experiment of mixing water and carbon in a known volume and measuring the resultant density). Again, the table shows (as Case 2) the results of the model for the equivalent dose rate.

The assumption that the radioactivity is uniformly mixed within the cylinder is an oversimplification, as the distribution of radioactivity is higher near the entrance to the bed (assumed to be the top of the tank for this work) and diminishes with bed depth. In order to simulate this effect the cylinder was divided into five sections of equal height. The radioactivity in each section was uniform, but the assigned value for each section decreased exponentially from top to bottom according to the following relationship

$$C(z) = C(Rn) \exp(-K_{SS}V/Q) \qquad\qquad \text{E-1}$$

where $C(z)$ is the concentration at depth z in the bed, $C(Rn)$ is the input concentration of the radon in the water, K_{SS} is the adsorption/decay constant (GAC- and water-specific; see Appendix C), V is the volume of the GAC and Q is the water flow rate. The resulting absorbed dose rate for this 'five-cylinder' model is shown in table E.1 as Case 3. As can be seen, the estimated dose rate is smaller for the two low-flow GAC units, compared with the single, well-mixed cylinder results. However, for the two examples of the high-flow case, the resulting dose rates are about the same.

Finally, calculations were also done with the CARBDOSE model (Rydell and Keene, 1993), which is intended for POE-type units only. Two calculations were done. The first assumes that the radioactive materials are confined to a point source; this yields an estimated equivalent dose of 0.148 µSv/h. The second is based on an extended radioactive source and gives an estimated equivalent dose of 0.173 µSv/h. These results are not very different and are essentially consistent with the results of Cases 1 and 2 shown in table E.1.

F

EPA Approach to Analyzing
Uncertainty and Variability

In the EPA (1995) risk assessment, the risk posed by exposure to radon by each pathway (that is, ingestion, inhalation of radon, and inhalation of progeny) is calculated by multiplying a series of terms that describe the link between radon concentration in water and the risk to the population using that water. The terms in this link include the radon concentration in water, human exposure levels per unit concentration, radiation dose per unit exposure, and cancer risk per unit radiation dose. On the basis of the best available data, EPA developed for each of these terms probability distributions that account for the value range of a parameter and the likelihood of exceeding any value within that range. Both uncertainty and variability are accounted for in these distributions. Monte Carlo simulations were used to propagate variance in the product of the terms.

UNCERTAINTY AND VARIABILITY IN THE RISK POSED BY INGESTED RADON GAS

In the EPA (1995) analysis, a cancer death risk was calculated for ingestion of radon gas in drinking water with a Monte Carlo analysis and the following formula:

$$R = (C) \ (F) \ (V) \ \ (365 \ \text{d/y}) \ (RF) \qquad \text{F.1}$$

where

R = risk of fatal cancer (per person per year) associated with ingestion of radon gas in water.

C = concentration of radon in water, pCi/L.
F = fraction of radon remaining in water at time of ingestion.
V = volume of water ingested, L/d.
RF = ingestion risk factor for radon gas (cancer-death risk per person per picocurie ingested).

Table F.1 summarizes the probability distributions used to represent value ranges for the parameters of this model. Each parameter has a variability distribution that is defined by two parameters, such as the mean and standard deviation. The two parameters are drawn from two distributions that represent their uncertainty. The concentration of radon in water, C, was derived by EPA from the National Inorganics and Radionuclides survey (Longtin 1990). A population-weighted population density function (PDF) was developed that was fitted to a lognormal distribution having a geometric mean of 200 pCi/L and a GSD of 1.85. The uncertainty in this distribution was obtained by using the Student t-distribution and the inverse chi-squared distribution to simulate the resampling of mean values and standard deviations. A sample size of 10 was used to resample from the Student t and inverse chi-squared distributions. Even though some 1,000 water systems are represented in the survey set, the assumed small sample size was selected to reflect the fact that the population-weighted concentration distribution was dominated by a small number of large water supplies for which there were a limited number of measurements. The variability of F, the fraction of radon remaining in water, was modeled as a beta distribution in the interval 0.1-0.3 with an uncertain mean and mode. The uncertainty of the mean was modeled as a uniform distribution in the range 0.7-0.9, and the uncertainty in the mode was modeled as a uniform distribution between 0.5 and the sampled mean value.

UNCERTAINTY AND VARIABILITY IN RISK POSED BY INHALED RADON GAS

In the EPA uncertainty analysis, the basic equation used to calculate inhalation uptake of radon gas is based on both the uncertainty and the variability of the unit dose factor, but the risk factor per unit dose is based on a single value. The unit dose factor for gas released from water includes three factors:

$$UD = (TF) \ (BR) \ (OF) \ (365 \ \mathrm{d/y}) \qquad \text{F.2}$$

where

UD = unit dose (pCi inhaled per year per pCi/L of radon in water).
TF = transfer factor, which is the increase in radon concentration in indoor air per unit radon concentration in water (pCi/L[air] per pCi/L[water]).
BR = breathing rate (L/d).
OF = occupancy factor (fraction of time person spends indoors).

TABLE F.1 Probability Density Functions Used in the Calculation of Risk of Cancer Posed by Ingestion of Radon Gas in Water (following notation of EPA 1995)

Parameter	Variability		Uncertainty	
	Distribution Type	Distribution Values	Distribution Type	Distribution Values
C (concentration of radon in water, pCi/L[water])	Lognormal distribution, $LN(\mu_\ell,\sigma_\ell)$	$C = LN(\mu_\ell, \sigma_\ell)$ $\mu_\ell = TS(m,s,n)$ $(\sigma_\ell)^2 = IChi(s,n)$	Student t distribution, $TS(m,s,n)$ Inverse chi-squared distribution, $IChi(s,n)$	$n= 10$ $m = \ln(200)$ $s = \ln(1.85)$
V (volume of water ingested, L/d)	Lognormal distribution, $LN(\mu_\ell,\sigma_\ell)$	$V = LN(\mu_\ell,\sigma_\ell)$ $\mu_\ell = TS(m,s,n)$ $(\sigma_\ell)^2 = IChi(s,n)$	Student t distribution, $TS(m,s,n)$ Inverse chi-squared distribution, $IChi(s,n)$	$n= 100$ $m = \ln(0.526)$ $s = \ln(1.922)$
F (fraction remaining)	Beta distribution, $B(m,md,min,max)$	$F=B(m,md,min,max)$ $min=0.5$ $max=1$ $m=U(a,b)$ $md=U(m,max)$ or $U(min,m)$	Uniform distributions $U(a,b)$ $U(m,max)$	$a=0.7$ $b= 0.9$ $min = 0.5$ $max=1$
RF (risk factor, cancer-death risk per person per pCi ingested)	This factor has uncertainty only		$RF=LN(\mu_\ell,\sigma_\ell)$	$\mu_\ell =$ $\ln(1.24 \times 10^{-11})$ $\sigma_\ell = \ln(2.42)$
Calculated individual risk	Uncertainty==> Variability	5th percentile	median	95th percentile
	5th percentile	1.7×10^{-8}	8.3×10^{-8}	3.4×10^{-7}
	mean	1.3×10^{-7}	6.2×10^{-7}	2.6×10^{-6}
	95th percentile	4.0×10^{-7}	1.9×10^{-6}	7.9×10^{-6}

m=mean value derived from a sample
s=standard deviation of a sample
n=sample size
md=mode of a sample
μ_ℓ=mean value of ln(x) in a lognormal distribution
σ_ℓ=standard deviation of ln(x) in a lognormal distribution

Table F.2 summarizes the probability distributions used to represent value ranges for the parameters of this model. Each parameter has a variability distribution that is defined by two parameters, such as the mean and standard deviation. They are drawn from distributions that represent the uncertainty of the two parameters. EPA used two-dimensional Monte Carlo simulations to develop an outcome distribution for UD. Two models were used to develop a distribution for the transfer factor (TF)—a one compartment indoor-air model and a three-compartment indoor-air model. Similar results were obtained from the two models. Because of the lack of information on the uncertainty and variability in the inhalation risk factor for radon gas, EPA calculated the mean population risk, PR, as

$$PR = (UD) \ (RF) \ (C_{\text{mean}}) \ (N) \qquad \text{F.3}$$

where

PR = population risk of fatal cancer (cancers per year) posed by ingestion of radon gas in water.

UD = unit dose (pCi inhaled per year per pCi/L of radon in water).

RF = risk factor, lifetime risk of cancer per person per pCi inhaled per year.

C_{mean} = population mean concentration of radon in water, pCi/L.

N = number of people in the population.

EPA used an inhalation risk factor, RF, of 1.1×10^{-12} cancer death per person per pCi/L of radon in water and a C_{mean} of 246 pCi/L. We can use this equation with $N = 1$ and $UD = 380$, which is the median value with respect to uncertainty and the mean value with respect to variability, to estimate the per caput risk within the exposed population. This gives a per caput risk of 1.0×10^{-7}, which is low compared with the mean individual risk (at median with respect to uncertainty) of 6.2×10^{-7} calculated for the ingestion pathway.

UNCERTAINTY AND VARIABILITY IN RISK POSED BY INHALED RADON PROGENY

The EPA (1995) report provided a separate calculation of variability and uncertainty of risk associated with exposure to radon progeny attributable to radon releases from household uses of water. In the EPA uncertainty analysis for radon progeny, the basic equation used to calculate risk is based on uncertainty and variability of the unit dose factor. The unit dose factor for radon progeny released from water includes three factors:

$$UD = TF \ [0.01 \ \text{WL/(pCi/L)} \ EF] \ (OF) \ [51.6 \ \text{WLM/(WL-yr)}] \qquad \text{F.4}$$

where

UD = unit dose (WLM per year per pCi/L).

TABLE F.2   Probability Density Functions Used by EPA (1995) in Calculation of Unit Dose of Radon Gas Inhaled after Transfer from Water (following notation of EPA 1995)

Parameter	Variability		Uncertainty	
	Distribution Type	Distribution Values	Distribution Type	Distribution Values
TF (transfer factor, pCi/L[air] per pCi/L[water])	Truncated lognormal $TLN(\mu_\ell, \sigma_\ell,$ min,max)	$TF = TLN(\mu_\ell, \sigma_\ell,$ min,max) $\mu_\ell = TS(m,s,n)$ $(\sigma_\ell)^2 = IChi(s,n)$ min=6. $\times$ 10^{-6} max=8. $\times$ 10^{-4}	Student t distribution, $TS(m,s,n)$ Inverse chi-squared distribution, $IChi(s,n)$	n= 25 m = $\ln(6.57 \times 10^{-5})$ s = $\ln(2.88)$
BR (breathing rate, L/day)	Truncated normal $TN(m,s,$min, max)	$BR = TN(\mu,\sigma,$ min.max) $\mu = TS(m,s,n)$ $\sigma^2 = IChi(s,n)$ min=3700 max=66,000	Student t distribution, $TS(m,s,n)$ Inverse chi-squared distribution, $IChi(s,n)$	n= 10 m = 13,000 s = 2880
OF (occupancy factor)	Beta distribution $B(m,md,min,$ max)	$OF=B(m,md,min,max)$ Min=0.17 Max=0.95 m=U(a,b) md=U(min,m) md=U(m,max)	Uniform distributions $U(a,b)$ $U(min,m)$ $U(m,max)$	a=0.65 b= 0.80 min = 0.17 max= 0.95
UD (unit dose, pCi inhaled per year per pCi/L)	Uncertainty==> Variability 5th percentile mean 95th percentile	5th percentile 17 32 250 800	median 57 380 1300	95th percentile 540 2000

m=mean value derived from a sample
s=standard deviation of a sample
n=sample size
md=mode of a sample
μ_ℓ=mean value of ln(x) in a lognormal distribution
σ_ℓ=standard deviation of ln(x) in a lognormal distribution
μ=mean value of x in a normal distribution
σ=standard deviation of x in a normal distribution

TF = transfer factor, which is the increase in radon concentration in indoor air per unit radon concentration in water [pCi/L(air) per pCi/L(water)].

EF = equilibrium factor (fraction of potential alpha energy of radon progeny that actually exists in indoor air compared with the maximum possible alpha energy under true equilibrium conditions).

OF = occupancy factor (fraction of time that person spends indoors).

From the *UD*, EPA calculated the unit risk factor, *UR* (lifetime risk of cancer death per pCi/L), and *IR*, the individual risk of cancer per individual:

$$UR = (UD)(RF) \qquad\qquad\text{F.5}$$

$$IR = (UR)\,C \qquad\qquad\text{F.6}$$

where

RF = risk factor, lifetime risk of cancer death per person-WLM of exposure.

C = concentration of radon in water, pCi/L.

Table F.3 summarizes the probability distributions used to represent value ranges for the parameters of this model. Each parameter has a variability distribution that is defined by two parameters, such as the mean and standard deviation. They are drawn from other distributions that represent the uncertainty. EPA again used two-dimensional Monte Carlo simulations to develop an outcome distribution for *UD*. As for the radon-gas inhalation model, both a one-compartment and a three-compartment model of indoor air were used, and they produced similar results for the transfer factor.

Table F.4 summarizes the calculation of PDFs used by EPA (1995) in the calculation of unit dose of radon progeny inhaled after transfer from water. From this table, we observe that the lifetime mean individual risk (at the median with respect to uncertainty) is 1.3×10^{-6} per year of exposure for the inhalation of radon progeny from water use. That is about twice the risk calculated for the ingestion pathway, 6.2×10^{-7}, and 10 times the mean (median) risk for inhalation of radon gas, 1.0×10^{-7}.

TABLE F.3   Probability Density Functions Used by EPA (1995) in Calculation of Unit Dose of Radon Progeny Inhaled after Transfer from Water (following notation of EPA 1995)

Parameter	Variability		Uncertainty	
	Distribution Type	Distribution Values	Distribution Type	Distribution Values
TF (transfer factor, pCi/L[air] per pCi/L[water])	Truncated lognormal $TLN(\mu_\ell, \sigma_\ell, min,max)$	$TF = TLN(\mu_\ell, \sigma_\ell, min,max)$ $\mu_\ell = TS(m,s,n)$ $(\sigma_\ell)^2=IChi(s,n)$ $min=6. \times 10^{-6}$ $max=8. \times 10^{-4}$	Student t distribution, $TS(m,s,n)$ Inverse chi-squared distribution, $IChi(s,n)$	$n= 25$ $m =$ $ln(6.57 \times 10^{-5})$ $s = ln(2.88)$
EF (equilibrium factor)	Beta distribution $B(m,md,min,max)$	$EF=B(m,md,min,max)$ $min=0.1$ $max=0.9$ $m=U(a,b)$ $md=U(min,m)$ $md=U(m,max)$	Uniform distributions $U(a,b)$ $U(min,m)$ $U(m,max)$	$a=0.35$ $b= 0.55$ $min = 0.1$ $max=.09$
OF (occupancy factor)	Beta distribution $B(m,md,min,max)$	$OF=B(m,md,min,max)$ $min=0.17$ $max=0.95$ $m=U(a,b)$ $md=U(min,m)$ $md=U(m,max)$	Uniform distributions $U(a,b)$ $U(min,m)$ $U(m,max)$	$a=0.65$ $b= 0.80$ $min = 0.17$ $max= 0.95$
UD (unit dose, WLM/y per pCi/L)	Uncertainty==> Variability 5th percentile mean 95th percentile	5th percentile 6.5×10^{-7} 1.2×10^{-5} 3.9×10^{-5}	median 1.2×10^{-6} 1.8×10^{-5} 6.4×10^{-5}	95th percentile 2.1×10^{-6} 2.7×10^{-5} 1.0×10^{-4}

m=mean value derived from a sample
s=standard deviation of a sample
n=sample size
md=mode of a sample
μ_ℓ=mean value of ln(x) in a lognormal distribution,
σ_ℓ=standard deviation of ln(x) in a lognormal distribution

TABLE F.4   Probability Density Functions Developed by EPA (1995) in Calculation of Risk, Unit Risk and Variation of Individual Risk Posed by Radon Progeny Inhaled after Transfer from Water (following notation of EPA 1995)

RF (risk factor, lifetime risk of cancer death per person-WLM)	This factor has uncertainty only		$RF = LN(\mu_\ell, \sigma_\ell)$	$\mu_\ell = \ln(2.83 \times 10^{-4})$ $\sigma_\ell = \ln(1.53)$	
UR (unit risk factor, lifetime risk of cancer death per pCi/L)	Uncertainty==> Variability	5th percentile	median		95th percentile
	5th percentile	1.3×10^{-10}	3.4×10^{-10}		8.9×10^{-10}
	mean	2.1×10^{-9}	5.1×10^{-9}		1.2×10^{-8}
	95th percentile	7.2×10^{-9}	1.8×10^{-8}		4.2×10^{-8}
C (concentration of radon in the water, pCi/L [water])	Lognormal distribution $LN(\mu_\ell, \sigma_\ell)$	$C = LN(\mu l, \sigma l,)$ $\mu_\ell = TS(m,s,n)$ $(\sigma_\ell)^2 = IChi(s,n)$	Student t distribution, $TS(m,s,n)$ Inverse chi-squared distribution, $IChi(s,n)$	$n = 10$ $m = \ln(200)$ $s = \ln(1.85)$	
IR (individual risk, lifetime risk of cancers per person-year)	Uncertainty==> Variability	5th percentile	median		95th percentile
	5th percentile	1.8×10^{-8}	5.4×10^{-8}		1.5×10^{-7}
	mean	5.2×10^{-7}	1.3×10^{-6}		3.2×10^{-6}
	95th percentile	1.9×10^{-6}	5.0×10^{-6}		1.3×10^{-5}

Index

Regulatory measures, *see* Legislation;
　　Standards
Relative risk models, 78, 81, 94-104
　　(passim), 137, 138, 140
　　excess relative risk, 96-97, 101
Reproductive effects, 12, 17, 30, 81
　　embryos, 81, 116
Residential Energy Consumption Study, 54-
　　55
Residential exposure, *see* Indoor radiation
　　exposure
Retinal cancer, 118, 119
Reverse osmosis, 177
Risk assessments, 7, 11, 12, 13-19, 30, 93-
　　104, 124-140, 198-199
　　absolute risk, 78, 93-94, 95, 223
　　atom bomb survivors, 28, 77-78, 80-81,
　　　95, 131, 137
　　attributable risk, 16, 100, 135, 223, 263
　　BEIR IV, 17, 93, 95-96, 99, 103
　　BEIR VI, 13, 15, 16, 17, 83, 93, 94, 95,
　　　99-100, 103, 122, 131, 134-135
　　biological basis, general, 13-14, 72-73,
　　　76-81, 122-123
　　cell-based, 122-123
　　disinfection and disinfection by-products,
　　　11, 20, 165-166, 171-178, 179,
　　　182, 195, 224, 254-256
　　　trihalomethanes (THMs), 165, 166,
　　　　179, 229, 254
　　EPA, 11, 14-15, 17, 28-30, 76, 126, 127-
　　　129, 132, 133-134, 137, 139, 140,
　　　162, 165-166, 168
　　　uncertainty, 127-129, 139, 140, 163,
　　　　260-267
　　exposure-age-concentration model, 135
　　exposure-age-duration model, 98, 135
　　groundwater, 11
　　heat-recovery ventilation, 149, 151, 158
　　ingestion risk, general, 14-15, 76-81, 182-
　　　183, 198, 260-261
　　inhalation risk, general, 15, 93-104, 182-
　　　183, 198, 261-267
　　IRCP, 94, 95
　　lung cancer, 1, 3, 6, 15, 16, 17, 78, 79, 80,
　　　82-83, 89-91, 93-104, 134, 166,
　　　182, 183
　　miners, 9, 15, 82-83, 93, 95-105 (passim)
　　　[ALL]
　　mortality, 1, 5-6, 8, 16, 18, 76-77, 78,
　　　100, 102

multimedia, 125 [???]
NCI, 93, 96-99, 101, 103
NCRP, 93-94
NIH, 17, 94, 100
occupational exposure, 63-64, 82, 93, 95,
　　100, 108
　　granular activated carbon controls,
　　　168, 169-171, 178, 179, 257-259
　　population attributable risk, 100, 135
　　process, steps in, 124-125
　　public education and, 31
　　relative risk models, 78, 81, 94-104
　　　(passim), 137, 138, 140
　　　excess relative risk, 96-97, 101
　　sensitivity analysis, 126-127, 130, 136
　　smoking, 2-3, 5-6, 8, 15, 16, 94, 95-96,
　　　98, 99, 100, 101, 103, 104, 193
　　soil and rock process, general, 3, 11, 29-
　　　30, 199
　　special populations, 81
　　stomach cancer, 6, 7, 15, 16, 72-73
　　water use and, general, 56, 84
　　see also Absolute risk; Attributable risk;
　　　Lifetime risk; Public education;
　　　Uncertainty
Risk management (mitigation), 4, 6, 11-12,
　　19-22, 30, 31, 144-197
　　action levels, 3, 172, 190
　　active subslab depressurization, 146-148,
　　　151-152, 153, 154, 157, 158, 223
　　aeration, 19-20, 161-167, 176-177, 249-
　　　251
　　airborne radiation, 15, 0, 19, 21-22, 31,
　　　144-160, 162-165, 182-188, 195
　　basements, 141, 142, 145, 146, 147, 148-
　　　149, 152, 155, 156, 157
　　concentration reduction, general, 144-145,
　　　148-151
　　construction materials and practices, 151-
　　　157
　　costs, 9-10, 20-21, 22, 160, 167, 174-175,
　　　183, 184, 187, 188, 192-193, 194,
　　　197
　　disinfection and disinfection by-products,
　　　11, 20, 165-166, 171-178, 179,
　　　182, 195, 224, 254-256
　　　trihalomethanes (THMs), 165, 166,
　　　　179, 229, 254
　　EPA, 4, 10, 11, 20-21, 150, 153, 154-155,
　　　156, 158-163, 165, 167, 168, 169,
　　　172, 177